Recent Advances in Wavelet Transforms and Their Applications

Edited by Francisco Bulnes

Published in London, United Kingdom

Recent Advances in Wavelet Transforms and Their Applications
http://dx.doi.org/10.5772/intechopen.97926
Edited by Francisco Bulnes

Contributors
Hitoshi Tsunashima, Masashi Takikawa, Wen-Liang Hwang, Jinn Ho, Shih-Shuo Tung, Meriane Brahim, Rahmouni Salah, Tifouti Issam, Claude Martias, Anthony Y. Aidoo, Gloria A. Botchway, Matilda A.S.A. Wilson, Bingcheng Li, Himanshu Monga, Dikshant Gautam, Saksham Katwal, Francisco Bulnes, Zion Menachem

First published in London, United Kingdom, 2022 by IntechOpen
IntechOpen is the global imprint of INTECHOPEN LIMITED, registered in England and Wales, registration number: 11086078, 5 Princes Gate Court, London, SW7 2QJ, United Kingdom

British Library Cataloguing-in-Publication Data
A catalogue record for this book is available from the British Library

Additional hard and PDF copies can be obtained from orders@intechopen.com

Recent Advances in Wavelet Transforms and Their Applications
Edited by Francisco Bulnes
p. cm.
Print ISBN 978-1-80355-738-0
Online ISBN 978-1-80355-739-7
eBook (PDF) ISBN 978-1-80355-740-3

Meet the editor

Dr. Francisco Bulnes obtained a Ph.D. in Mathematical Sciences from the Mathematics Institute, National Autonomous University of Mexico (IM-UNAM). He also has two post-doctorate degrees in Mathematics, obtained in Cube and Russia. He is currently the director of Investigación Internacional Avanzada en Matemáticas e Ingeniería (IINAMEI), Mexico. He is editor-in-chief and reviewer of national and international journals of mathematics and a member of various international committees of science. He is head of the Research Department, Grupo de Investigación del Tecnológico de Estudios Superiores Chalco (GI-TESCHA). Dr. Bulnes has published more than 120 papers in mathematics and physics research journals. He has also written several books in these areas and has many theories, theorems, and math objects to his credit. He is a recipient of various honors and awards from universities and governmental and non-governmental organizations. He is a distinguished member of the Czech Republic Mathematics Society (JCFM).

Contents

Preface

Over the last few years, the wavelet transforms in their multiple forms—continuous, discrete, real, and complex—have revolutionized the way we manage spectral and computing data for multi-resolution analysis, graph representation in trees, sub-representation of spectral bands in wavelet decomposition to an image, time series data to localization of time-meaning signals, spectrometry, and spectrograms data, and more. It can be used in numerous ways, which is attributable to its ability to complement Fourier transforms data in its spectral image while also providing a few advantages over Fourier transforms in terms of reduced calculations when analyzing particular frequencies. While it has many applications in Fourier analysis and dynamical systems, its main strengths lie in the interpretation of spectra in two-dimensional spectral images—searching for signals of a known, non-sinusoidal shape. The wavelets play a significant role in the typical STFT/Morlet analysis. From a purely mathematical perspective, a wavelet series is a representation of a square-integrable (real- or complex-valued) function by a certain orthonormal series created by wavelets. Notably, The Riesz theorem and other features of functional analysis about the convergence of various wavelet series that can be created after the generation of an appropriate wavelet are crucial to the consistency of these representations. However, this does not come without its fair share of challenges. We have observed a few problems with the wavelet transform, along with other functional transforms: the Mexican hat wavelet, Haar Wavelet, Daubechies wavelet, triangular wavelet, and many more.

Dr. Francisco Bulnes
Professor,
IINAMEI Director,
Head of Research Department in Mathematics and Engineering,
Tescha, Mexico

Section 1

Introduction to Analyisis and Signal Processing

Chapter 1

Analysis and Processing of One-Dimensional Signals Using Wavelet

Meriane Brahim, Rahmouni Salah and Tifouti Issam

Abstract

In recent years, wavelet analysis has become an effective and important computational tool in signal processing and image processing applications. Wavelet analysis is known for its successful approach to solving the problem of signal analysis in both the time domain and frequency domain. The analysis of the nonstationary signal generated by physical phenomena has posed a great challenge for various conversion techniques. Transformation techniques such as Fourier transform (FT) and short Fourier transform (STFT) fail to analyze nonstationary signals. But wavelet transform (WT) techniques may be able to efficiently analyze both stable and unstable signals. WT is able to analyze one-dimensional signals, such as audio signals and two-dimensional signals such as images. In this chapter, we discuss wavelet transduction techniques and their applications in detail and focus on the analysis and processing of the wave-encoded laser signal as one-dimensional electrical signals and its use in alarm systems. In the second stage, we filter the speech signal and determine the fundamental frequencies using wavelet transformation.

Keywords: wavelet transform, nonstationary signal denoising, lasers sources, alarm system, discrete wavelet transforms (DWT)

1. Introduction

This chapter introduces the study and realization of a laser barrier alarm system, after the laser is obtained by an electronic device, the wireless control system is connected to the control room to announce the application in real-time, the laser is used in many application fields, from industry to medicine, it uses an alarm system to detect and deter intruders. Basic security includes protecting the perimeter of a military base or a safe distance in unsafe places or near a government place. The first stage secures surrounding access points such as doors and windows; the second stage consists of internal detection with motion detectors that monitor movements, there are several types of products on the market, and the system you buy can be wired or wireless. Wired systems use cables to connect each device to the central control panel. A wireless system that runs on batteries and transmits its signals at a radio frequency, and there are no cables. In this chapter, we rely on the realization of a coded laser barrier that is sent between two units, processing the signal, and comparing the agreed conditions, and in order to be high accuracy, we suggest that we use wavelet transduction to process the received signal and know the frequencies that achieve the activation of the alarm.

2. Warning and protection systems

Alarm and protection systems have been developed in many fields, in many areas such as the military field, where lasers have been used for detection. Any attempt to break through the wall of the military barracks. In addition to the technologies in this field, we are working on the use of coded lasers, which means that we send pulses that are very limited in frequency and periodicity, as well as in the number of pulses during a pre-agreed period of time [1].

Laser pulses can be obtained through an electronic circuit with analog processing, and to eliminate any noise in the receiving circuit, we filter the signal using wavelet transformation, thanks to which we get high accuracy and an effective system that works in real-time [2].

In addition to activating the alarm, this system can also work to send information via radio waves to the control room so that the leadership can make decisions at the same time [2, 3].

2.1 Photovoltaic barriers

Photovoltaic barriers are optical or electronic systems consisting of a sensor (receiver) and a light source (emitter). The light source can be an ordinary lamp, an infrared emitter (e.g., a pulse), LEDs, or a laser emitter (**Figure 1**) [2].

2.2 Laser barrier application

Single barriers consist of a separate interacting transmitter and receiver. Reflex barriers and detectors combine sensor and light source in a single box. In reflex

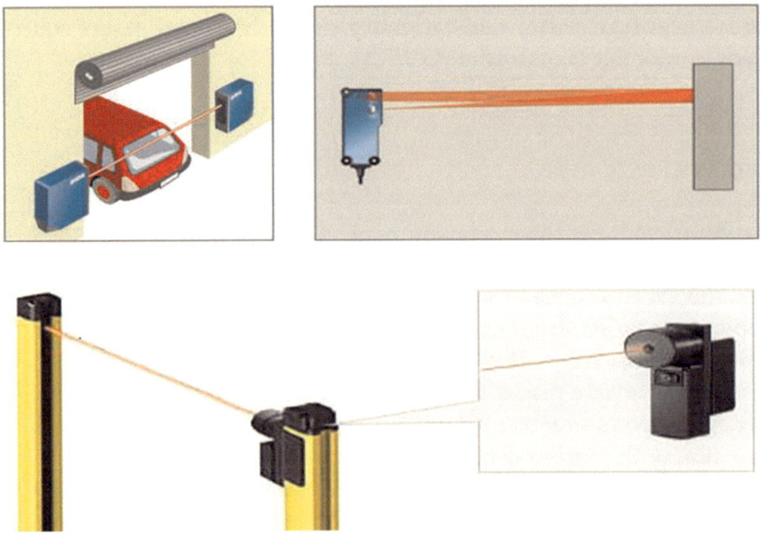

Figure 1.
Photovoltaic barriers.

barriers, the emitted light beam is returned by a reflector (prism, reflective sheet) to the receiver [4]. Light barriers register an interruption in the light beam and convert the information. If an object passes through the beam of an optoelectronic barrier, the sensor generates a predefined electrical output signal. It triggers an alarm. Detectors send a very fine infrared beam and react to the reflection of light from an object. The maximum detection distance depends largely on the reflectance rate, shape, color, and surface quality of the material [5–7].

3. System structure

The corresponding figure shows the stages of transmitting and processing information that determines with high accuracy all the electronic circuits on which this project depends, as it consists of a laser transmitter encoded between two transmitting and receiving units, the processing stage, and the activation of the alarm with the radio wave communication system (**Figure 2**).

3.1 Laser transmitter circuit

The corresponding circuit shows the electronic card responsible for producing the laser pulses, with the possibility of changing the frequency and the periodic ratio [8] (**Figure 3**).

Circuit diagram simulated in Crocodile Technology 607 (**Figure 4**).

3.2 Basic Astable 555 oscillator circuit

The 555 IC can be used to create a free-running as table oscillator to continuously produce square wave pulses (**Figure 5**).

The previous electronic circuit generates square signals or pulses and this is related to the values of the resistors and capacitors and depends on connecting the second electrode with the sixth and separated between them by a special resistance for discharging the capacitor through the seventh electrode.

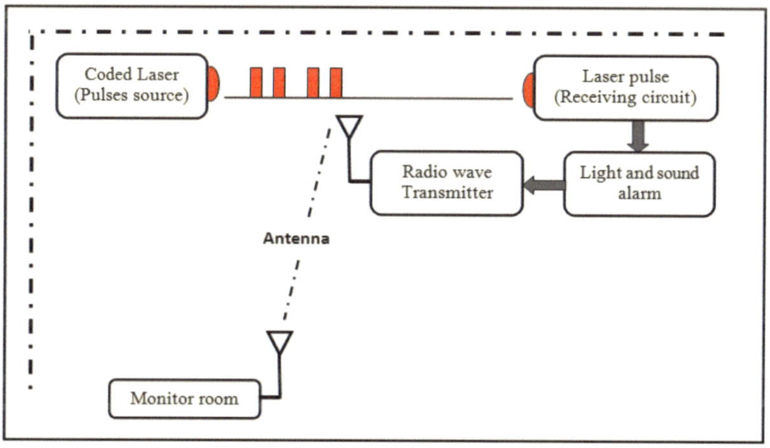

Figure 2.
Configuration of the whole system.

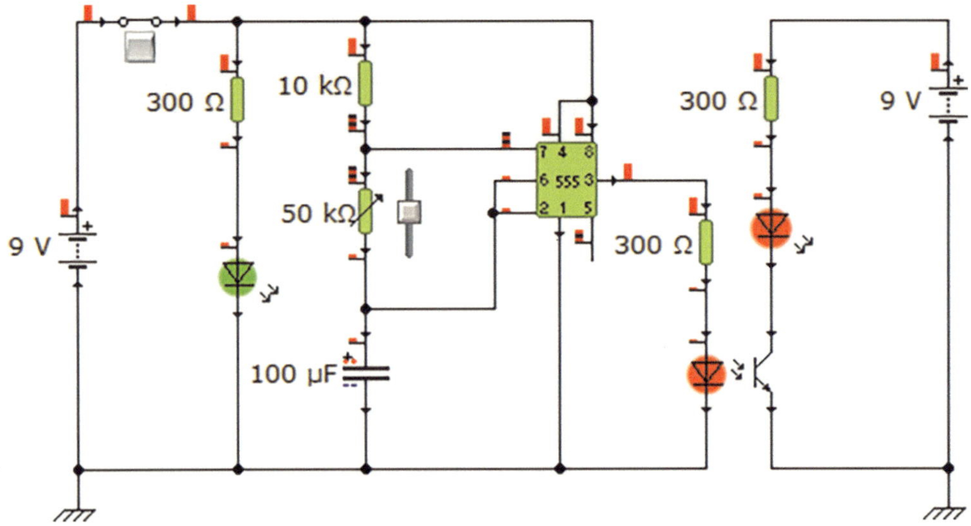

Figure 3.
Laser transmission circuit.

In the initial conditions, the tension between the poles of the capacitor is equal to zero, and when the generator is connected, the capacitor starts charging in an exponential equation until it reaches the value 2/3 Vcc. Then, the outlet voltage ceases. In these conditions, the capacitor starts discharging until it reaches 1/3 Vcc, and this is repeated. The process is repeated several times and this is called an unstable oscillator.

$$T = t_1 + t_2 \quad (1)$$

$$t_1 = 0.69(\ R_1 + R_2\ )C \quad (2)$$

$$t_2 = 0.69.R_2.C \quad (3)$$

The output frequency of oscillations can be found by inverting the equation above for the total cycle time giving a final equation for the output frequency of an Astable 555 Oscillator as:

$$F = \frac{1}{T} \quad (4)$$

Applied results of the transmission circuit obtained using an oscilloscope (**Figure 6**).

The output signal can be controlled by connecting a direct polarizing diode between the sixth and seventh poles so that we can determine the charging constant and the discharging constant, which in turn controls the type of output signal (**Figure 7**).

3.3 Laser receiver circuit

One of the advantages of the practical amplifier called IC LM 741 is the comparison between the inverting and non-inverting input signals so that the output voltage is symmetrical and varies according to the comparison process, and in this case, it works in the nonlinear characteristic.

DOI: http://dx.doi.org/10.5772/intechopen.104107

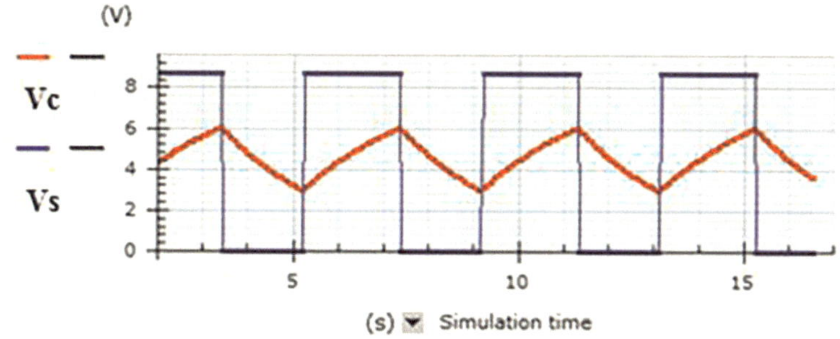

(a)

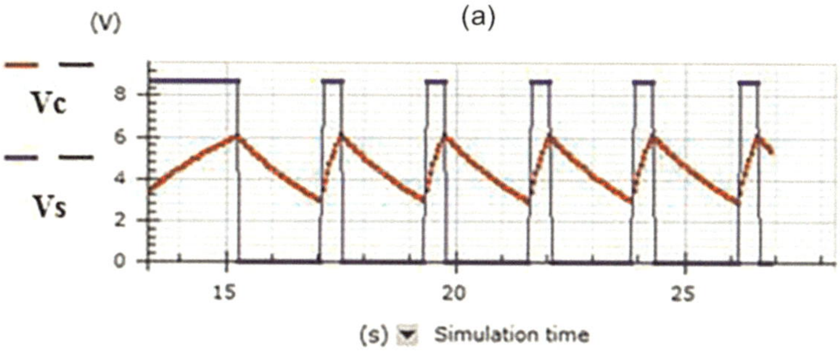

(b)

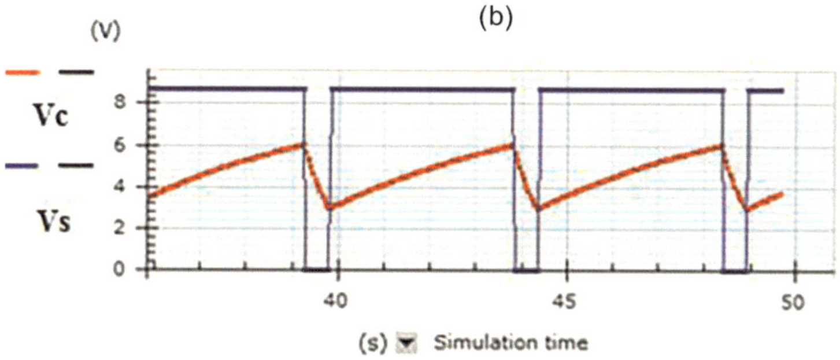

(c)

Figure 4.
Simulation results of the laser transmitter circuit: (a) square electric signal, (b) pulse signal, and (c) rectangular signal.

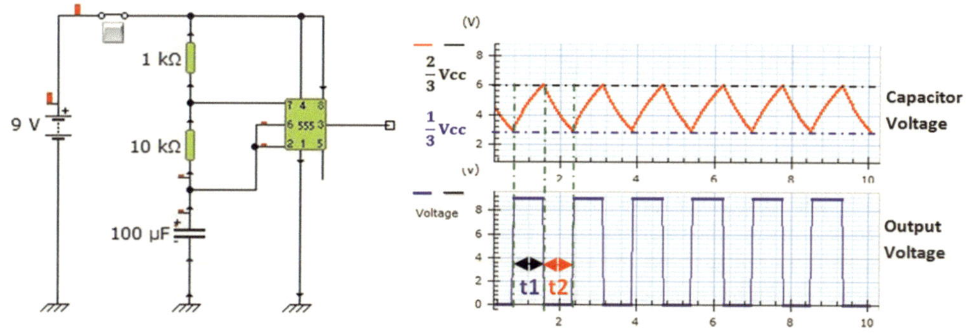

Figure 5.
555 oscillator.

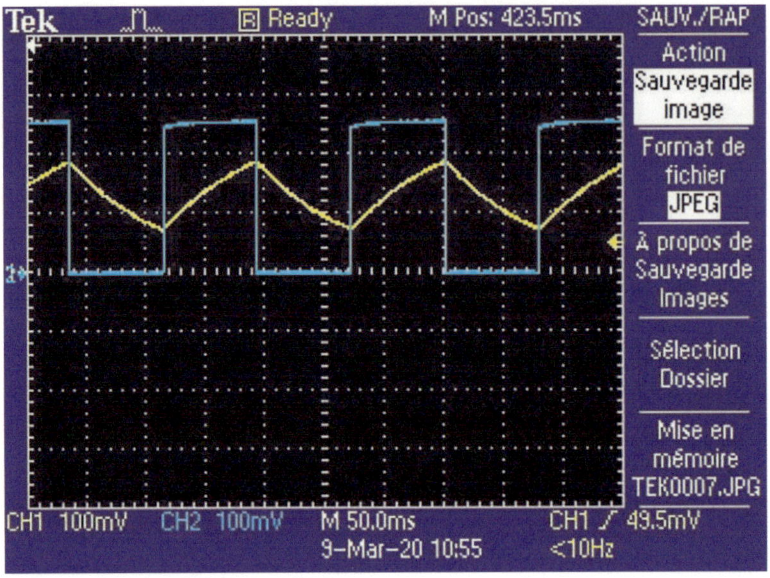

Figure 6.
The real results of the laser transmitter circuit—square electric signal.

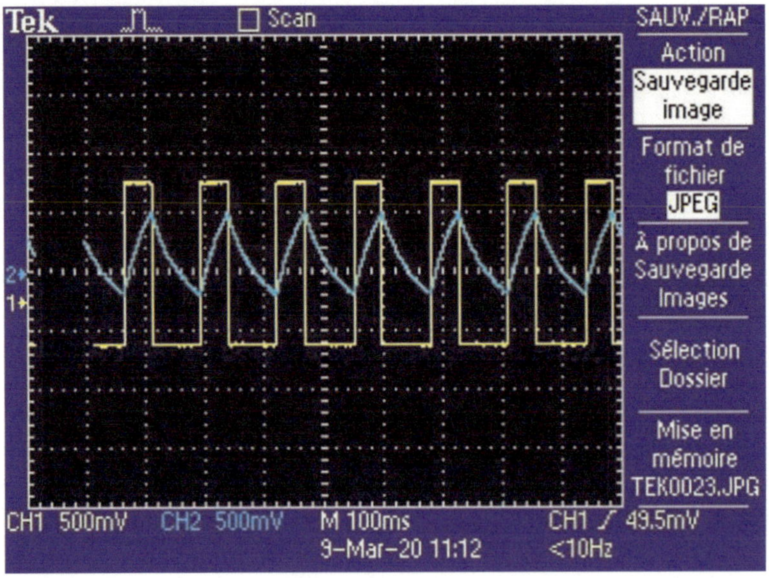

Figure 7.
Transmitter circuit results—pulsed signal.

The practical amplifier may be used in the process of amplifying weak signals if it operates in the linear characteristic so that the relationship between the input signal and the output signal is fixed (**Figure 8**).

After receiving and processing the laser beam, we get the following signal (**Figure 9**).

The signal obtained is either square, rectangular or pulsed, depending on the transmission signal, and accordingly, the average value can be calculated as follows:

$$\overline{E} = \frac{1}{T}\int_{0}^{T} E(t).dt \tag{5}$$

$$\overline{E} = \frac{1}{T}\int_{0}^{t1} E(t).dt + 0 \tag{6}$$

$$\overline{E} = \frac{1}{T}E.t_1 \tag{7}$$

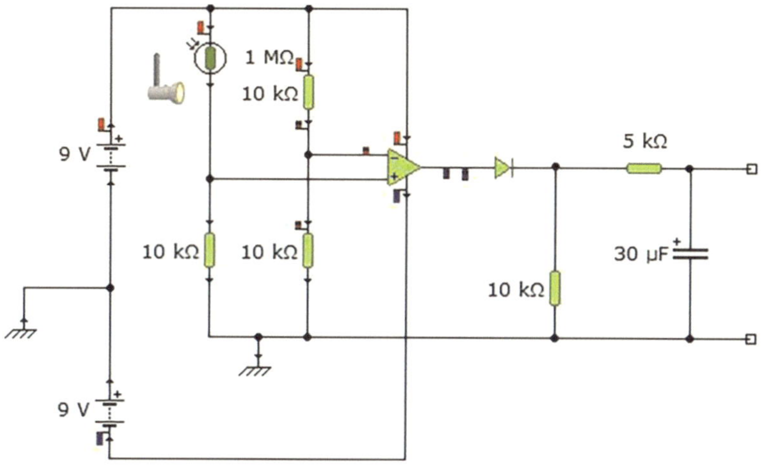

Figure 8.
Laser receiving circuit using comparator LM 741.

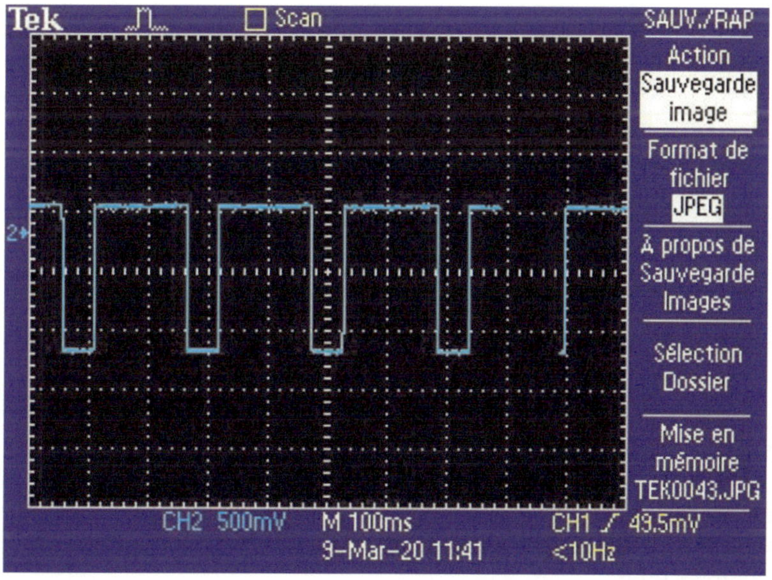

Figure 9.
Laser beam reception signal.

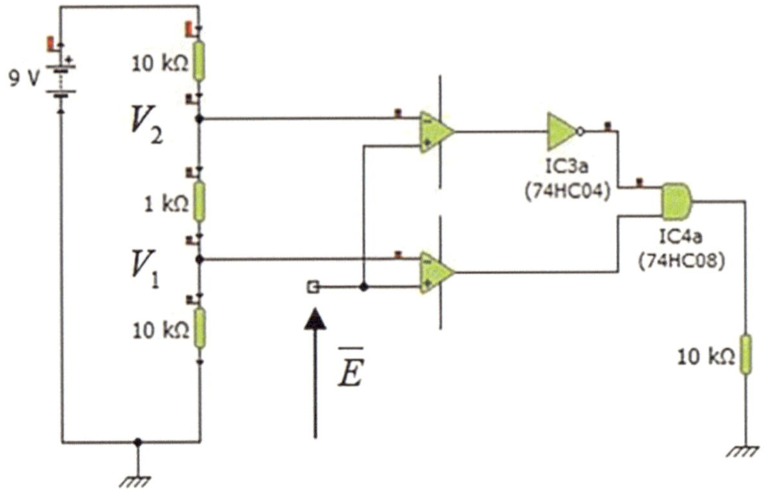

Figure 10.
Comparison circuit with two amplifiers LM 741.

Knowing that the mean value is proportional to the duration t_1 then

$$\overline{E} = \eta.E \tag{8}$$

η, the cyclic report.

The following circuit is to compare the average value of the main signal and the reference voltage, to increase the accuracy by adding another practical amplifier and some electronic components (**Figure 10**).

$$\overline{E} = 2.5V, V1 = 2.3V, V2 = 2.8V$$

So $V_1 < \overline{E} < V_2$ for this condition the alarm system is in the off state because no laser beam cut between the two cards (transmission and reception).

3.4 Processing circuit

The main processing circuit consists of the following electronic components (**Figure 11**).

The corresponding figure represents the printed circuit of the project using Express PCB software and electronic components CMS (**Figure 12**).

The following figure shows the real picture of the project with the processing circuit and data transmission using radio waves (**Figure 13**).

3.5 Experimental results

The figures above show the transmit and receive signal with duty cycles of approximately 25 and 75%, respectively (**Figures 14** and **15**).

—Emission signal

—Reception signal +/− 12

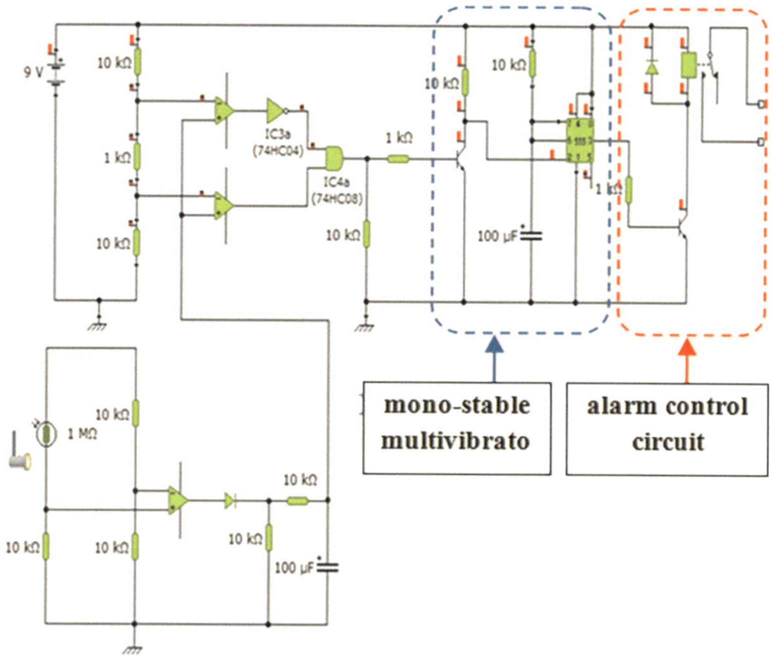

Figure 11.
Electronic circuit of the project.

4. Analysis of the reception signal by wavelet transformation

In recent years, wavelet transform (WT) has been relied upon as a radical alternative to signal processing, especially after the discovery of the problematic Fourier theory, which focuses on the frequency domain, and given that most physical quantities are unstable here. It can be said that we need to develop methods of processing and analysis, it can be said that in the seventies of the last century the wavelet theory was used in several areas, such as noise removal, image improvement, classification of audio signals, etc. [9]. Chu and Kim applied the Morlet wavelet transform to analyze the effect of noise.

The wavelet theory depends on important properties in processing so that the value of a window of the mother wave is proportional to the frequency of the signal to be processed. There are two types of wavelet transports: continuous (CWT) and discrete (DWT) existing transport processes. Both transformations are continuous in time (analog), and with their help, analog signals can be represented [10].

4.1 General theory of CWT

In this work, we only touched upon some of the basic equations, definitions, and concepts of the wavelet transform, and a more rigorous mathematical treatment of this topic can be found in [10, 11]. The time-continuous wavelet transform of $f(t)$ is defined as:

$$CWT_{\psi}f(a,b) = \frac{1}{\sqrt{a}} \int_{-\infty}^{+\infty} f(t).*\psi\left(\frac{t-b}{a}\right).dt \tag{9}$$

where:

- $CWT_{\psi}f(a,b)$ is the wavelet coefficient of the function $f(t)$
- $\Psi(t)$ is the analyzing wavelet
- a (> 0) is the scale parameter
- b is the position parameter

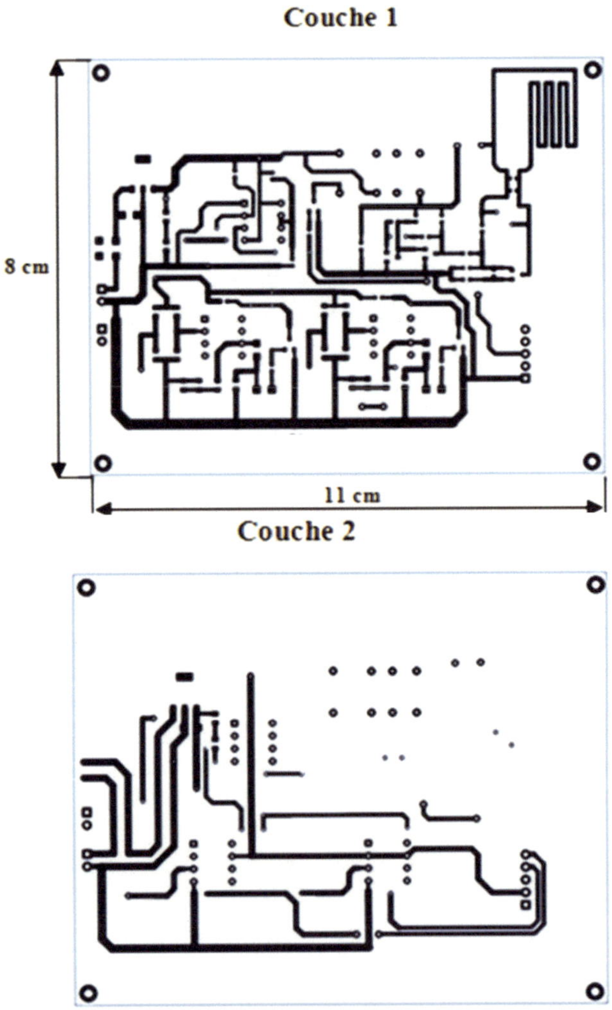

Figure 12.
The mother card for the project.

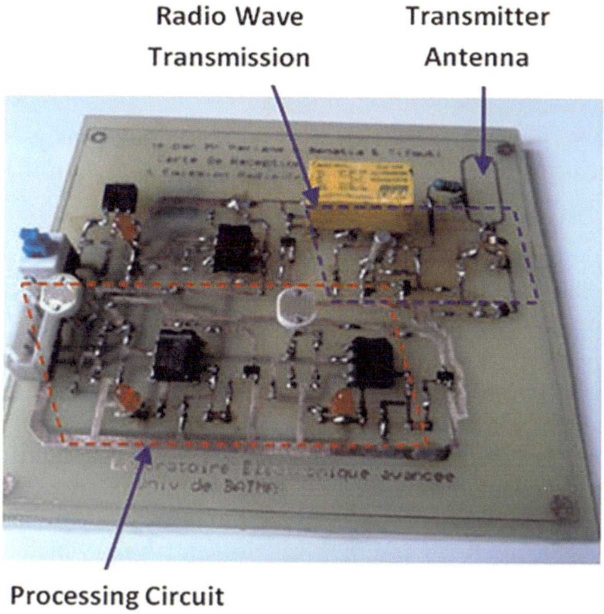

Figure 13.
The final electronic card for the project.

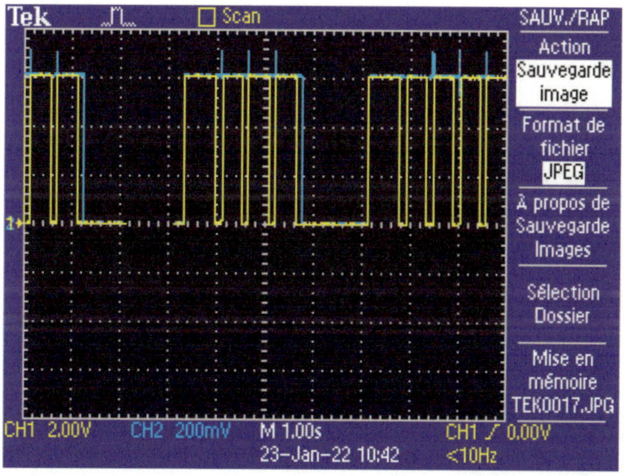

Figure 14.
Experimental results for a pulsed signal.

4.2 Temporal and spectral resolutions in the CWT

The signal processing in the time and frequency domain is very important so that we can know the different frequency components, where we express the time resolution in the time domain σt, and the spectral resolution in the frequency domain σw from CWT as:

$$\sigma_t^2(\gamma) = \int_{-\infty}^{+\infty} (t - u\gamma)^2 |\varphi_\gamma(t)|^2 dt \quad (10)$$

Wavelets perform and deliver a high-accuracy scale-based analysis of specific data [12]. They then find a wide range of applications and uses for these waves including signal processing, mathematics, and numerical analysis, and for their good use in signal and image processing, they are a radical alternative to the fast Fourier transform where the DWT provides a time-frequency representation when a non-stationary tool is needed for processing and analysis, DWT can be used. The study showed that discrete wave delivery has a high performance in processing speech signals so far [9, 13].

A computer-assisted experiment (CAM) is not fundamentally different from a laboratory experiment as it was traditionally performed using different measuring instruments and laboratory equipment, but the computer integration in the processing of the laser pulse receiving signal brings several advantages. The data acquisition process can be automated, and the measurement results can be saved and processed easily and in a very short time by various software tools. In addition, the presentation of the results in graphic form is greatly simplified and their scientific analysis, which facilitates the analysis and use of the wavelet transform on the obtained signal.

4.3 Acquisition of reception signal with CoolEdit

CoolEdit program is used to record electrical signals that are proportional to the physical quantities to be processed and will be recorded in a one-dimensional matrix at a sampling frequency of 64 kHz using a 16 transducer called the sampling period.

$$T_e = \frac{1}{F_e} = \frac{1}{64 \times 10^3} = 0.015 \times 10^{-3} s \quad (11)$$

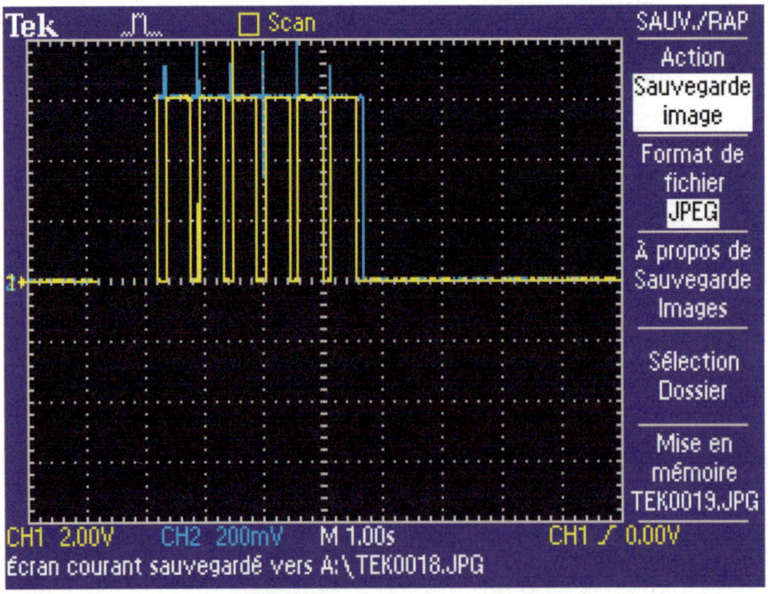

Figure 15.
Experimental results of a rectangular signal.

The following design represents how to record the electrical signal from the receiving circuit using computer-assisted experiments (**Figure 16**).

The signal obtained is considered to be a non-stationary signal and it is also made up of several signals which cannot be recorded using the oscilloscope because it is technically unable to track and oscillate instantaneous signals of very high frequency. Response time, although it is very infinite, also take a slow-motion until it stabilizes, and this is called differential vibration.

The molar figure shows the difference between the signal recorded by the cathode oscilloscope and the one recorded by a computer-assisted experiment (**Figures 17** and **18**).

In the normal case, the signals resulting from the receiving circuit can be drawn using the cathode oscilloscope, but it does not give us rapid changes that can only be detected by using alternative devices. Computer-supported experiments were selected and then we process the data, which is represented in one-dimensional arrays using the MATLAB program. For these calculations we used the MATLAB cwt function.

After processing the stored matrix, we get the following figure, after applying the wavelet transform with the selection of the Haar wavelet algorithm to analyze the signal into several levels, where we can know the basic frequencies of the received laser signal (**Figure 19**).

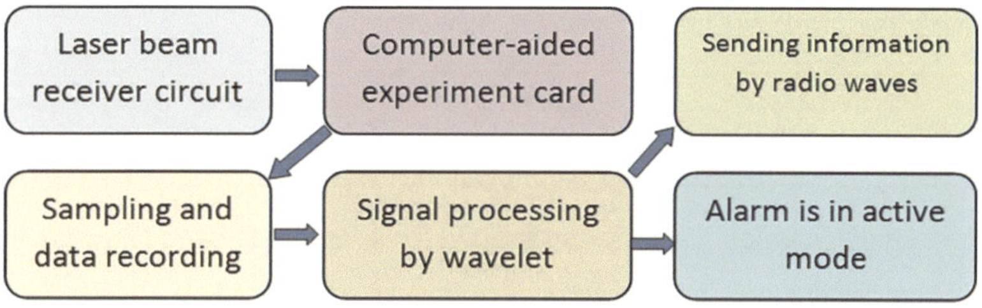

Figure 16.
Diagram of the proposed method.

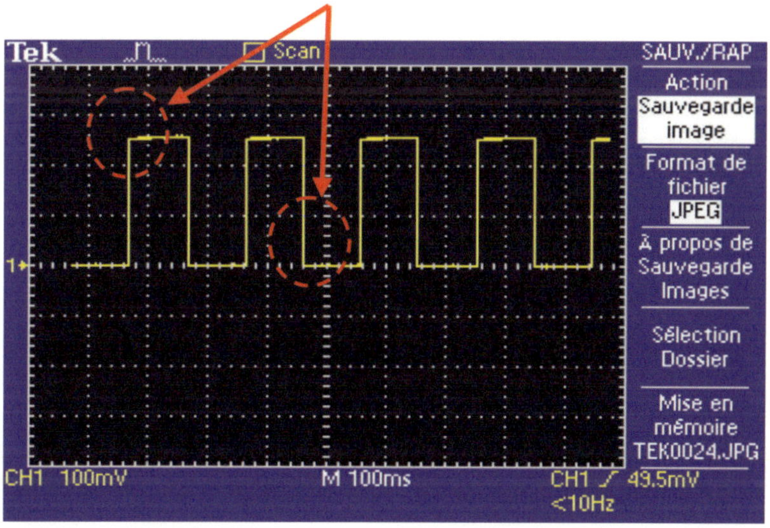

Figure 17.
The results of the transmitter circuit by the oscilloscope.

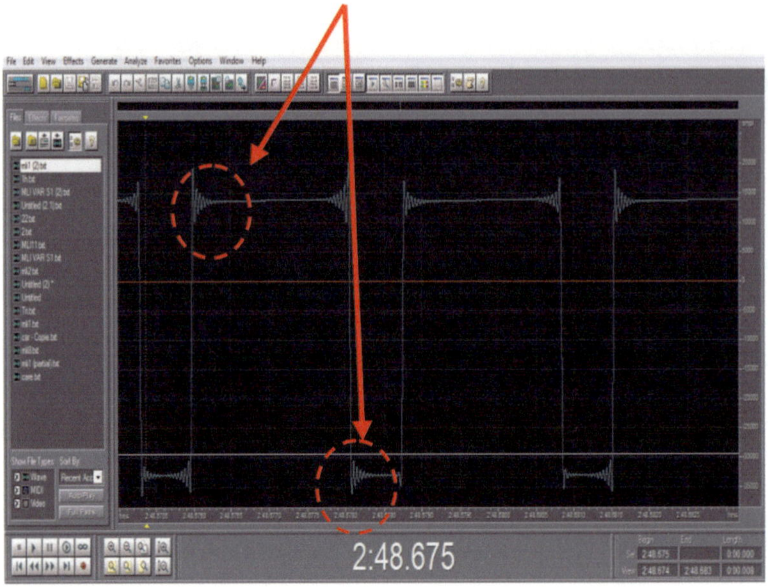

Figure 18.
Real results of the transmitter circuit by CoolEdit.

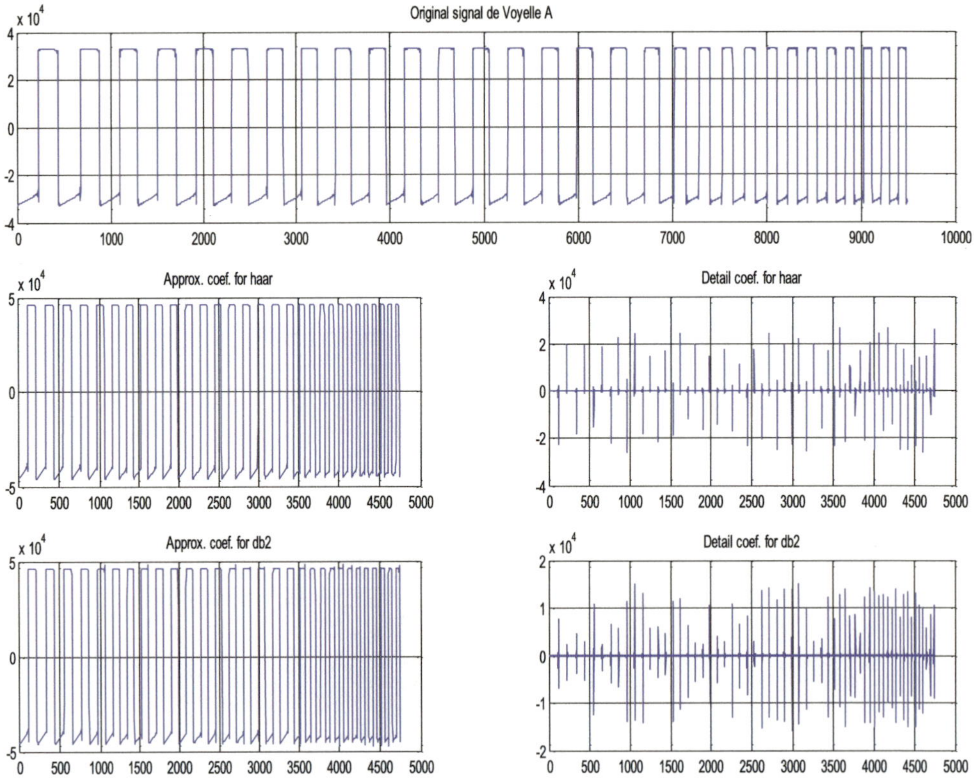

Figure 19.
Simulation results of wavelet transformation on the receiving signal at an increased frequency.

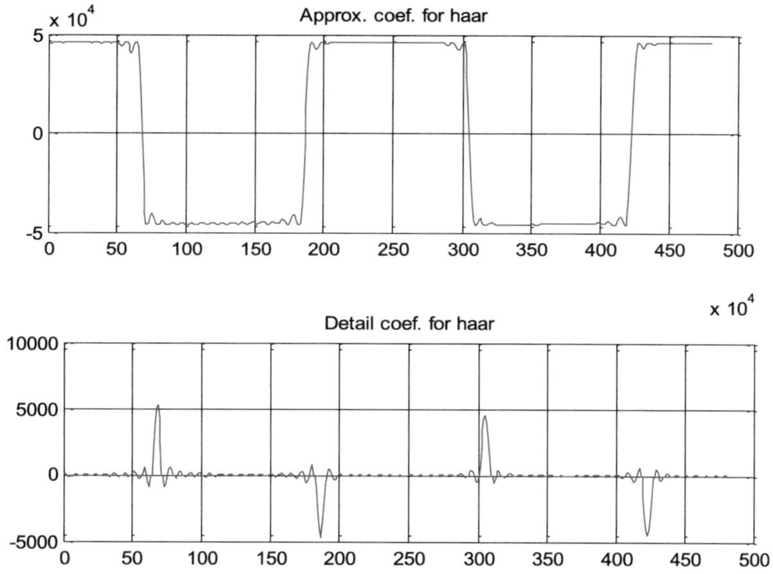

Figure 20.
The Haar wavelet transform on the received signal.

The following figure shows how to deconstruct the signal obtained in the receiving circuit, showing the frequency resulting from the differential vibration, which we would not have obtained without applying the wavelet theory (**Figure 20**).

5. Denoising and enhancement speech signal using wavelet

After the development of the field of information processing and the discovery of wavelet analysis, it became common to process signals and unstable physical quantities, such as speech analysis, and sound signature discovery by recognizing the basic frequencies of letters. Wavelets proved successful in processing such signals, which is an alternative to all methods. It was there before where it processes data in real-time using time-wave resolution, waveforms rely on the Henning window [7]. Recognition performance depends on frequency domain coverage. The goal of good speech recognition is to increase the bandwidth of the wavelength without significantly affecting the time accuracy. This can be done by collecting the white noise of the wave, which is difficult to detect and remove by traditional methods.

5.1 Speech enhancement methods

There are many methods available to improve speech, reduce noise, and the quality of audio signals, and each algorithm has a principle that it depends on in the processing methodology, and this depends on the goals we want to reach. In this chapter, we propose the corresponding layout, which enables us to filter the audio signals and identify the frequencies that make up them, as shown in **Figures 2** and **21**.

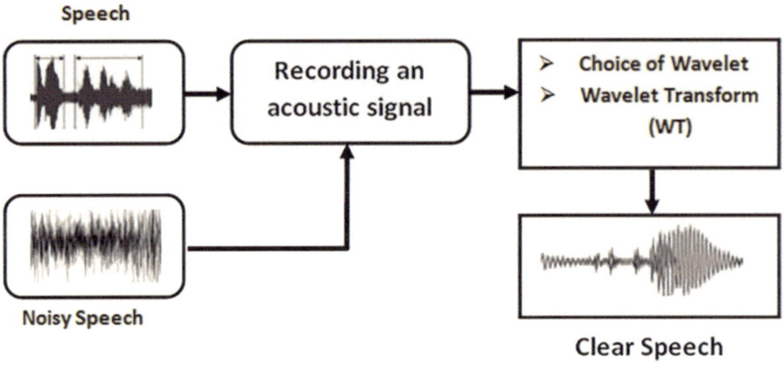

Figure 21.
Speech enhancement method.

5.2 Detection of singularity of the impulse noise signal in CWT

For noise evaluation, the oscillation of the acoustic signals is regarded as a considerable important metric. The CWT is often applied to detect the singularities of a transient signal (**Figures 22** and **23**; **Table 1**).

In the general case, we can rely on three main frequencies to define letters and words, after removing noise and applying wavelet transform. Practically, the third frequency can be neglected because it may be close in several letters, and we are

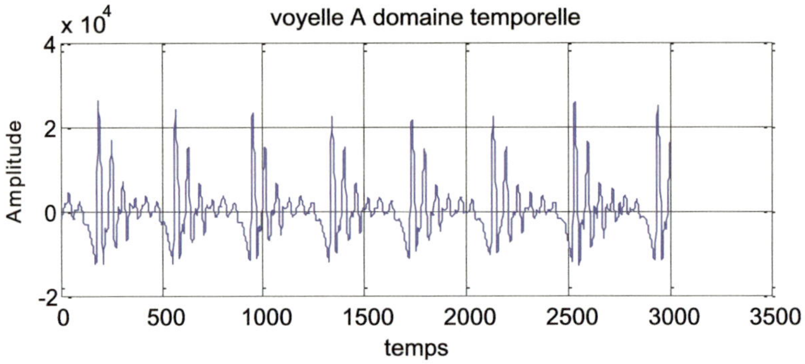

Figure 22.
Vowel A time domain.

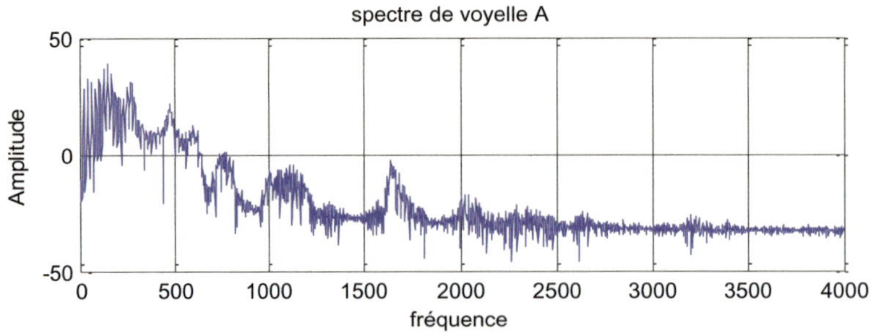

Figure 23.
The vowel A with the spectrum and the CWT.

Frequency Hz Vowel	Frequency F1	Frequency F2	Frequency F3
A	625	1491	2356
E	387	1985	2875
I	246	2018	3196
U	312	750	2079
O	313	756	2271

Table 1.
Shows the various frequencies for the vowel "A, E, I, U, O."

satisfied with only the first and second frequencies, especially if the noise is removed at an acceptable rate.

The figure opposite shows the letter E after noise cancelation (**Figure 24**).

5.3 Enhancement vowel "A," by the wavelets

The following figure shows the resulting multi-resolution vowel "A," by the wavelets (**Figure 25**).

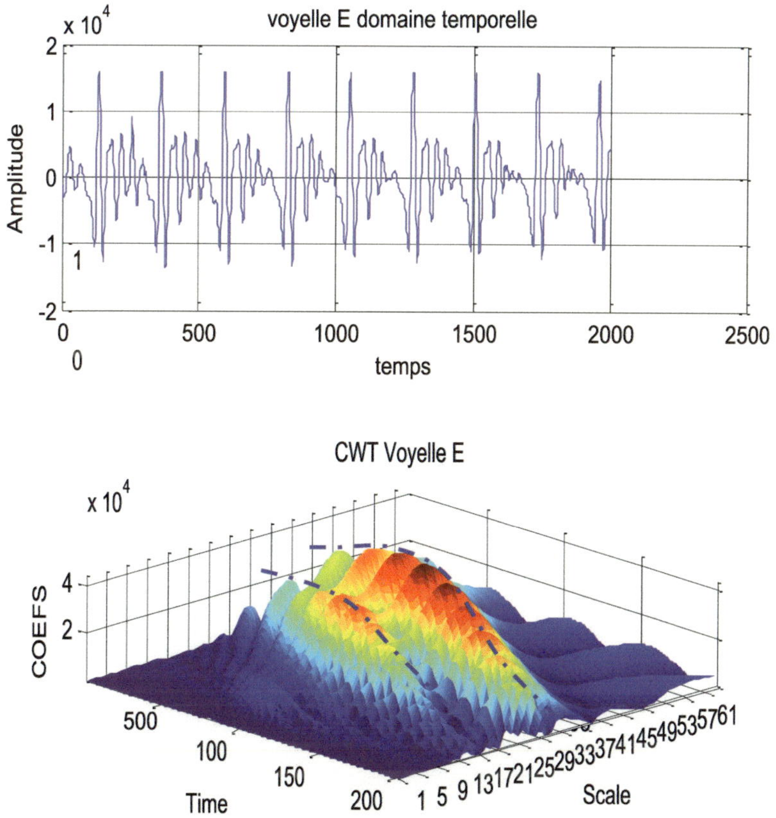

Figure 24.
The vowel E with the spectrum and the CWT.

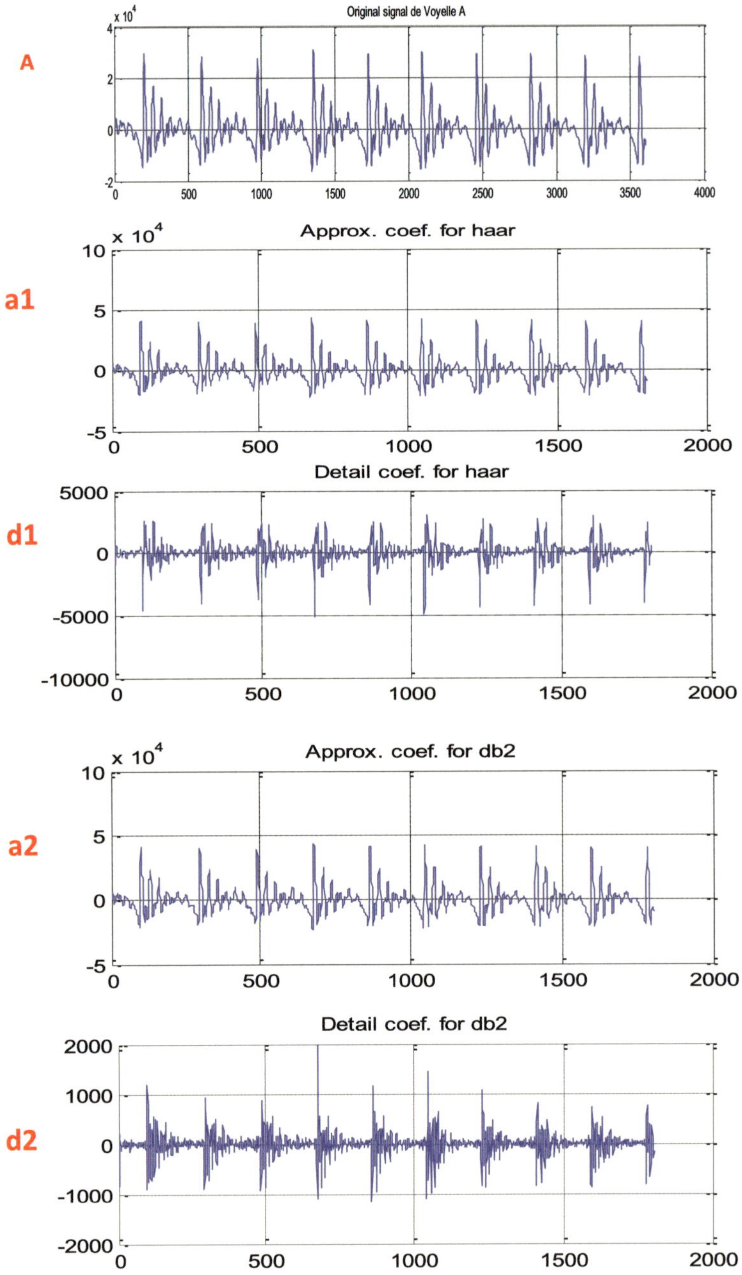

Figure 25.
Vowel "A," levels of decomposition by the wavelets: (A) original speech signal, (a1) appro 1, (d1) level 1, (a2) appro 2, and (d2) level 2. A = a1 + d1 + a2 + d2

6. Conclusion

In this work, we relied on processing the data sent between a transmitter and receiver, which were coded laser pulses that control a sophisticated warning system that can be used in a military field. In order for this system to be effective, we decided to analyze the received signal by converting waves that have proven successful in several fields. After deconstructing the received signal using wavelet transformation, it can be said that the pulses are not as perfect as we expected, because in fact they consist of several signals with different frequencies, and this leads to the lack of dependence on this signal in controlling any system, especially if it requires high accuracy or efficiency in Performance such as controlling the speed of the DC motor or transmitting digital information by wireless communications, especially if it is a high frequency.

The use of wavelet transform is not limited to electrical signal processing but can be applied to improve sound quality, and this method depends on the signal threshold that each waveform parameter of the signal is compared to a certain threshold. Using wavelet transform to remove noise from a signal may require identifying which components contain the noise, and then reconstructing the signal without those components after eliminating them.

In contrast to the STFT transformation which has constant accuracy at all times and at all frequencies, we can say that WT has good temporal accuracy and low-frequency accuracy at high frequencies, and good frequency accuracy and low temporal accuracy at low frequencies.

Wavelet decomposition is very similar to the Gabor transform: the speech signal is written as a superposition of the displacement and expanding waves. The main frequencies can be recognized in a short time and with great accuracy especially in word recognition programs if we get rid of the noise in the audio signal and this is what we discussed in this section where the wavelet transform can be used to reduce noise.

After applying the proposed algorithm, noise from speech signals was successfully reduced by using wavelet transform. This method gives us a hands-on approach on how to filter sound and recognize different frequencies.

Author details

Meriane Brahim*, Rahmouni Salah and Tifouti Issam
Higher School of Technological Education Skikda ENSET, Azzaba, Algeria

*Address all correspondence to: tlcom_brahim@yahoo.fr

References

[1] Arifin MAZ. Automation Security System with Laser Lights Alarm on Web Pages and Mobile Apps. London: IEEE; 2019. DOI: 10.1109/ISCAIE.2019.8743998

[2] ASecureLife. Best Home Security Systems. 2019. Available from: https://www.asecurelife.com/best-home-security-system/ [Accessed: January 7, 2019]

[3] Tseloni A, Thompson R, Grove L, Tilley N, Farrell G. The effectiveness of burglary security devices. Security Journal. 2017;**30**:646-664. DOI: 10.1057/sj.2014.30

[4] Rai A. Low Cost Laser Light Security System in Smart Home. London: IEEE; 2019. DOI: 10.1109/CISCT46613.2019.9008141

[5] Arianne B. Design and implementation of an Arduino-based security system using laser light. LPU-Laguna Journal of Engineering and Computer Studies. 2019;**4**(2):1

[6] Rahmouni S, Zighed L, Chaguetmi S, Daoudi M, Khelifa M, Karyaoui M, et al. Correlation between photoluminescence and ellipsometric measurements of porous silicon layers. The Journal of Optoelectronics and Advanced Materials. 2018;**12**(9-10):553-558

[7] Guo R, Song L. Space Laser Chaotic Security System. London: IEEE; 2019. DOI: 10.1109/ICCSEC.2017.8446898

[8] Abrar MM. Design and implementation of astable multivibrator using 555 timer. IOSR Journal of Electrical and Electronics Engineering (IOSR-JEEE). 2017;**12**(1):22-29

[9] Torkamani R, Sadeghzadeh RA. Wavelet-based Bayesian algorithm for distributed compressed sensing. Journal of Information Systems and Telecommunication (JIST). 2019;7(2):87. DOI: 10.7508/jist.2019.02.001

[10] Meriane B. Denoising and enhancement speech signal using wavelet. Journal of Information Systems and Telecommunication (JIST). 2021;**9**(1):37-44. DOI: 10.52547/jist.9.33.37

[11] Kahrizi MR. Long-term spectral pseudo-entropy (LTSPE): A new robust feature for speech activity detection. Journal of Information Systems and Telecommunication (JIST). 2018;**6**(4):204-208

[12] Kim J. Time-frequency characterization of hand-transmitted, impulsive vibrations using analytic wavelet transform. Journal of Sound and Vibration. 2007;**308**:98-111. DOI: 10.1016/j.jsv.2007.07.046

[13] Singh P, Joshi SD, Patney RK, Saha K. The Fourier decomposition method for nonlinear and non-stationary time series analysis. Proceedings of the Royal Society A: Mathematical, Physical and Engineering Sciences. 2017;**473**:20160871

Chapter 2

Time Frequency Analysis for Radio Frequency (RF) Signal Processing

Bingcheng Li

Abstract

In high-density radio frequency (RF) signal environments, receivers usually acquire signals from multiple sources. These RF signals may be co-channel and co-duration, which cause significant difficulties for processing them. Time-Frequency analysis combined with a projection pursuits graph approach provides an effective way to detect, separate, and classify these multiple source RF signals. Time-frequency analysis includes a spectrogram approach and a scalogram approach. The feature points on the instantaneous frequency function of a frequency modulation radio frequency (FMRF) signal can be extracted from either the spectrogram or scalogram of this FMRF signal. With the projection pursuits graph approach, these feature points are grouped into time-frequency functions to represent the multiple components for the separation, detection, and classification of this multisource FMRF signal.

Keywords: connected graph, time-frequency manifold, multisource signal separation, projection pursuits, spectrogram, wavelet, scalogram

1. Introduction

In congested electromagnetic environments, the radio frequency (RF) signals a receiver receives may include multiple time-frequency overlapped signals transmitted from multiple emitters. Traditional RF signal processing techniques may have difficulty separating and processing these multisource signals.

The instantaneous frequency function of an FMRF signal can be approximately modeled by low-order polynomials. Chirplet transforms and polynomial chirplet transforms have been investigated to process multisource FMRF signals [1–13]. These approaches separate and process multisource cochannel FMRF signals effectively; however, their implementations are expensive due to high dimensional transforms.

With a first-order polynomial approximation of the phase function of an FMRF signal, the short-time Fourier transform approach provides a simple and low-cost implementation for instantaneous frequency estimation. Unlike chirplet and polynomial chirplet transforms which need to perform transforms from time to high dimensional frequency and chirp spaces, the short-time Fourier transform approach creates spectrograms and only needs to perform time to frequency transforms. Using fast Fourier transforms, the short-time Fourier transform for a local window with size W only needs $O(W log W)$ computations, which are much lower than utilizing chirplet or polynomial chirplet transform approaches [14, 15].

Spectrograms are created by a fixed window size Fourier transform. For a low-frequency component, it needs a large window to capture enough changes for this low-frequency component. However, for a high-frequency component, it needs a small window to have a high time resolution. The constant window size for spectrogram cannot satisfy these conflict requirements. To address this issue, a natural extension is to perform Fourier transforms with changeable window sizes. For high-frequency components, small window sizes are used to perform transforms while large window sizes are used for low-frequency transforms. This extension leads to the wavelet transforming with constant weights in the window, creating a scalogram. The weight functions could also be other functions that lead to different wavelet transforms. For instance, choosing a Gaussian function creates Gabor or Morlet wavelet transform.

Spectrograms or scalograms provide the time-frequency representation of a multisource FMRF signal. Separating this multisource FMRF signal into each independent source component needs further processing. The ridge points of spectrograms or scalograms over some thresholds generate the points for instantaneous frequency functions. In this chapter, a connected graph will be introduced to extract instantaneous frequency functions when they are not crossed with each other. When the instantaneous frequency functions are crossed with each other, a projection-pursuit approach is described to separate and extract these instantaneous frequency functions.

2. A FMRF signal model and its spectrogram

In this section, an FMRF signal model with a single source is introduced, and a Sinc() function for its spectrogram is derived from this model with short-time Fourier transforms. Then, this FMRF signal model and its spectrogram computation are extended to multiple component FMRF signals.

2.1 An FMRF signal model with a single component

A single component FMRF signal is described by the following model,

$$s(t) = A_0 \exp\left(i\varphi(t)\right) + n(t) \tag{1}$$

where $\varphi(t)$ is the instantaneous phase function of this FMRF signal and $n(t)$ represents additive white noises.

Another function to describe an RF signal is its instantaneous frequency function $f(t)$. The frequency function $f(t)$ and phase function $\varphi(t)$ have differential and integral relations are shown as follows,

$$f(t) = \frac{d}{dt}\varphi(t) \text{ and } \varphi(t) = \varphi_0 + \int f(t)dt \tag{2}$$

It is shown in Eq. (2) that the frequency function $f(t)$ of an RF signal represents its phase function $\varphi(t)$ with only a constant phase uncertainty φ_0. Due to this reason, in this chapter, we will focus on applying the frequency functions to analyze and process FMRF signals.

The instantaneous frequency function $f(t)$ is a one-dimensional manifold imbedded into a two-dimensional time-frequency space. Without noises ($n(t) = 0$), a single component FMRF signal and its time-frequency manifold are shown in **Figure 1**, where the left side is an FMRF signal, and the right side is its time-frequency manifold.

Since the time-frequency manifold of an FMRF signal is the representation of this FMRF signal, we can use time-frequency manifolds to classify or recognize RF signals. Also, the time-frequency manifolds of an FMRF signal provide an estimation of its instantaneous frequencies.

2.2 Sinc() function of the time-frequency image or spectrogram of a single component FMRF signal

The time-frequency image or spectrogram of an FMRF signal $s(t)$ is the magnitude of the short-time Fourier transform of $s(t)$,

$$I(\omega, t) = \left| \sum_{\tau=0}^{W-1} A_0 \exp\left(i\varphi(t+\tau)\right) e^{-j\omega\tau} + n_I(\omega, t) \right| \tag{3}$$

where W is the window size of this short-time Fourier transform.

Expanding $\varphi(t+\tau)$ by its first-order Taylor series around t in a local window, we have

$$\varphi(t+\tau) = \varphi(t) + c_1(t)\tau + O\left(\tau^2\right) \tag{4}$$

where $c_1(t) = \frac{d\varphi(t)}{dt}$, which is the instantaneous frequency of $s(t)$ at time t, and the first-order Taylor expansion in (4) is a linear approximation to the phase function $\varphi(t)$ in its local window.

Under the linear approximation of a phase function, the time-frequency image or spectrogram of $s(t)$ can be derived from (3) and (4),

$$I(\omega, t) \approx \left| \sum_{\tau=0}^{W-1} A_0 \exp\left(i\varphi(t)\right) e^{j(c_1(t)-\omega)\tau} + n_I(\omega, t) \right|$$

$$= \left|A_0 \exp\left(i\varphi(t)\right)\right| \left| \sum_{\tau=0}^{W-1} e^{j(c_1(t)-\omega)\tau} + n_I(\omega, t) \right| = \left| \frac{A_0 \sin\left(\frac{W(c_1(t)-\omega)}{2}\right)}{\sin\left[(c_1(t)-\omega)/2\right]} + n_I(\omega, t) \right|$$

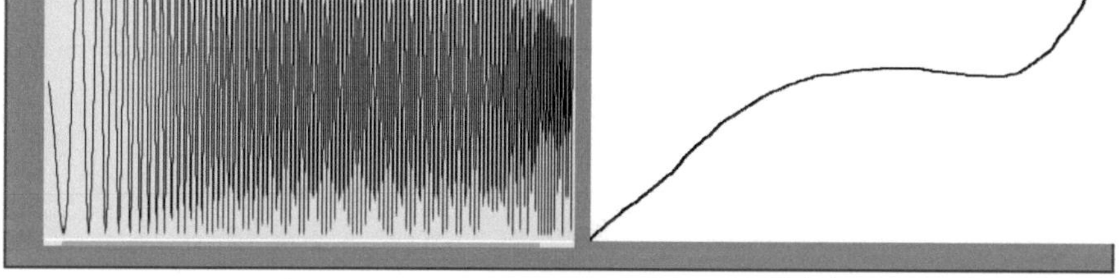

Figure 1.
An FMRF signal (left) and its time-frequency manifold (right).

Thus, we have approximated the spectrogram of an FMRF signal

$$I(\omega,t)\approx\left|\frac{A_0 \sin\left(W(c_1(t)-\omega)/2\right)}{\sin\left[(c_1(t)-\omega)/2\right]}+n_I(\omega,t)\right| \tag{5}$$

Eq. (5) shows that when noises $n_I(\omega,t)$ are low, at a given time t, the spectrogram of an FMRF signal is a *Sinc()* function in the frequency direction. This *Sinc*() function reaches its maximum at the instantaneous frequency $c_1(t)$.

The spectrogram of the FMRF signal in **Figure 1** is shown in **Figure 2**. **Figure 2** shows the *Sinc()* patterns in the vertical (frequency) direction.

2.3 Spectrograms of multisource and cochannel multisource and cochannel FMRF signals

The multisource and co-channel FMRF signals received by a receiver is modeled as

$$x(t) = n(t) + \sum_{k=1}^{K} s_k(t) = n(t) + \sum_{k=1}^{K} a_k \exp\left(i\varphi_k(t)\right) \tag{6}$$

where, K is the number of sources for the FMRF signal, and $n(t)$ is the noises of the receiver.

Similar to (4), a linear approximation in a local window is used to approximate the phases for the multisource FMRF signal,

$$\varphi_k(t+\tau) = \varphi_k(t) + c_{1,k}(t)\tau + O(\tau^2) \tag{7}$$

where $c_{1,k}(t)$ is the instantaneous frequency for the component signal $s_k(t)$.

An equation to compute the spectrogram for the multisource FMRF signal is derived by substituting (7) into (6),

$$I(\omega,t) = \left|\sum_{k=1}^{K}\frac{A_k \sin\left(\frac{W\left(c_{1,k}(t)-\omega\right)}{2}\right)}{\sin\left[(c_{1,k}(t)-\omega)/2\right]}+n_I(\omega,t)\right| \tag{8}$$

where, $n_I(\omega,t)$ is the spectrogram noises. (8) shows that when noises $n_I(\omega,t)$ are low, the spectrogram of this multisource FMRF signal is the magnitude of the summation of multiple *Sinc()* functions.

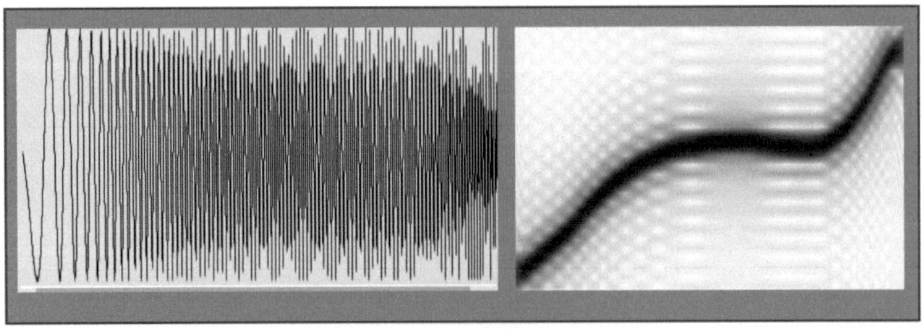

Figure 2.
An FMRF signal and its time-frequency image (spectrogram).

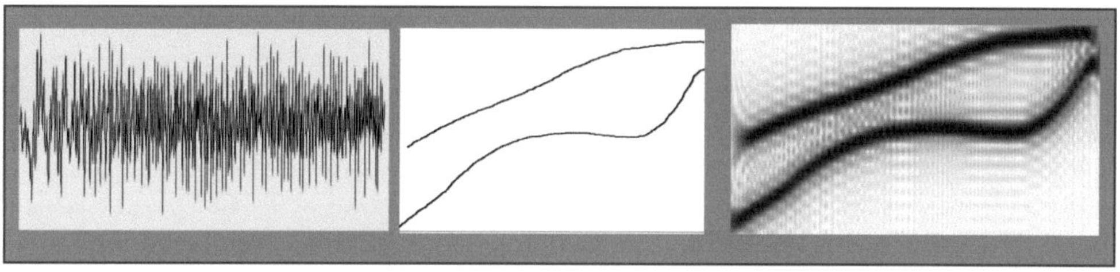

Figure 3.
A multisource and cochannel FMRF signal and its spectrogram. The left is the FMRF signal, the middle is its instantaneous frequency function, and the right is its spectrogram.

A multisource and cochannel FMRF signal is shown in **Figure 3**. The right side of **Figure 3** shows the spectrogram of this FMRF signal where the *Sinc()* function patterns are distributed in the frequency direction.

3. A scalogram as an extension of spectrogram

It is shown in Section 2 that the spectrogram of an FMRF signal created by short-time Fourier transform (STFT) demonstrates *Sinc()* patterns. In this section, we will show that the wavelet transforms with a uniform window is a direct extension of the STFT, so the scalogram generated by wavelet transforms is a direct extension of the spectrogram.

3.1 Scalogram computaion of a single component FMRF signal

Define a rectangle window function

$$R(t) = \begin{cases} 1, 0 \le t < 1 \\ 0, otherwise \end{cases} \tag{9}$$

For a window size W, we have

$$R\left(\frac{t}{W}\right) = \begin{cases} 1, 0 \le t < W \\ 0, otherwise \end{cases} \tag{10}$$

With the help of the window function $R\left(\frac{t}{W}\right)$, the STFT in (3) is written into the following format,

$$I(\omega, t) = \left| \sum_{\tau=-\infty}^{\infty} A_0 \exp\left(i\varphi(t+\tau)\right) R\left(\frac{\tau}{W}\right) e^{-j\omega\tau} + n_I(\omega, t) \right| \tag{11}$$

The computation of spectrograms in (11) is the same as that in (3). They both give the same STFT for spectrogram computations by a uniform distributed weight function $R\left(\frac{t}{W}\right)$ with a fixed window size W.

When the window size W in (11) is chosen to be changed by $W = \frac{c}{\omega}$ as frequency ω changes,

$$I(\omega,t) = \left| \sum_{\tau=-\infty}^{\infty} A_0 \exp\left(i\varphi(t+\tau)\right) R\left(\frac{\tau}{\frac{c}{\omega}}\right) e^{-j\omega\tau} + n_I(\omega,t) \right| \tag{12}$$

where c is a constant.

Eq. (12) is a wavelet transform with a mother wavelet $R\left(\frac{\tau}{c}\right)e^{-j\tau}$, and $\frac{1}{\omega}$ is the scale of the wavelet transform. A uniform distributed weight function is discussed in this chapter. $R()$ can be a different weight function. If $R()$ is chosen to be a Gaussian function, (12) will give Gabor or Morlet wavelet transform.

To distinguish scalogram from spectrogram, we change $I(\omega,t)$ to $W(\omega,t)$ to show that the scalogram is generated by a wavelet transform. The scalogram of an FMRF signal is calculated by the following equation.

$$W(\omega,t) = \left| \sum_{\tau=-\infty}^{\infty} A_0 \exp\left(i\varphi(t+\tau)\right) R\left(\frac{\omega\tau}{c}\right) e^{-j\omega\tau} + n_W(a,t) \right| \tag{13}$$

(11) and (13) show the close relationship between STFT and the wavelet transform. The scale in the wavelet transform is inversely proportional to the frequency while the scale STFT is fixed. In other words, the wavelet transform can be treated as an adaptive STFT where the window size of the STFT (referred to as scale in the wavelet transform) adapts to the frequency change of the STFT. When the frequency is high, the window size is small so as to catch the high resolution in time. When the frequency is low, the window size is large so as to obtain a high resolution in frequency. In this sense, a wavelet transform usually creates a higher performance than an STFT due to the wavelet's adaptive properties.

Similar to the derivation of the spectrogram calculation by summation in (3), the scalogram calculation can also be derived using wavelet transforms. Writing (13) into a summation format creates the following expression,

$$W(\omega,t) = \left| \sum_{\tau=0}^{\frac{a}{\omega}-1} A_0 \exp\left(i\varphi(t+\tau)\right) e^{-j\omega\tau} + n_I(\omega,t) \right| \tag{14}$$

The scalogram calculated by (13) is further simplified by substituting the FMRF signal of (4) into (14),

$$W(\omega,t) \approx \left| \sum_{\tau=0}^{\frac{a}{\omega}-1} A_0 e^{j\varphi(t)} e^{j(c_1(t)-\omega)\tau} + n_I(\omega,t) \right| \tag{15}$$

The $e^{j\varphi(t)}$ in (15) does not depend on τ. Thus, (15) can be written into another form,

$$W(\omega,t) = |e^{j\varphi(t)}| \left| A_0 \sum_{\tau=0}^{\frac{a}{\omega}-1} e^{j(c_1(t)-\omega)\tau} + e^{-j\varphi(t)} n_I(\omega,t) \right| = \left| A_0 \sum_{\tau=0}^{\frac{a}{\omega}-1} e^{j(c_1(t)-\omega)\tau} + e^{-j\varphi(t)} n_I(\omega,t) \right| \tag{16}$$

If noise term $e^{-j\varphi(t)} n_I(\omega,t)$ is denoted as $n_W(\omega,t)$, the wavelet transform in (16) becomes,

$$W(\omega,t) = |\frac{A_0 \sin\left(\frac{\alpha}{\omega}\frac{c_1(t)-\omega}{2}\right)}{\sin\left(\frac{c_1(t)-\omega}{2}\right)} + n_W(\omega,t)| \tag{17}$$

Eq. (17) shows that similar to the spectrogram $I(\omega,t)$, the scalogram $W(\omega,t)$ also demonstrates the *Sinc()* properties near the instantaneous frequency $c_1(t)$. The difference is that the *Sinc()* function of the spectrogram $I(\omega,t)$ oscillates in an equal period while the scalogram $W(\omega,t)$ oscillates in an increasing period as frequency increases.

The comparison between spectrogram and scalogram is shown in **Figure 4**. In **Figure 4**, the frequency of the FMRF signal is chosen as 10 kHz in the local window. For the spectrogram, the window size is chosen as 20. For the scalogram, the window size is selected to change from 18 to 22. At the center frequency 10 kHz, the mask size of the scalogram is the same as the window size for spectrogram 20. **Figure 4** shows that the *Sinc()* function oscillates with the same frequency in the frequency direction for spectrogram. However, the oscillation frequency for the scalogram increases from low frequency to high frequency.

3.2 Scalogram computaion of a multisource FMRF signal

Similar to the computation of a single source FMRF signal, the scalogram computation of a multisource FMRF signal is given by replacing the fixed-size window summation in (8) with the frequency-dependent window summation,

$$W(\omega,t) = |\sum_{k=1}^{K} \frac{A_k \sin\left(\frac{\alpha}{\omega}\frac{(c_{1,k}(t)-\omega)}{2}\right)}{\sin\left(\frac{(c_{1,k}(t)-\omega)}{2}\right)} + n_W(\omega,t)| \tag{18}$$

Eq. (18) shows that the scalogram of each component of a multisource FMRF signal is a *Sinc()* function in the local frequency direction, which is similar to the spectrogram.

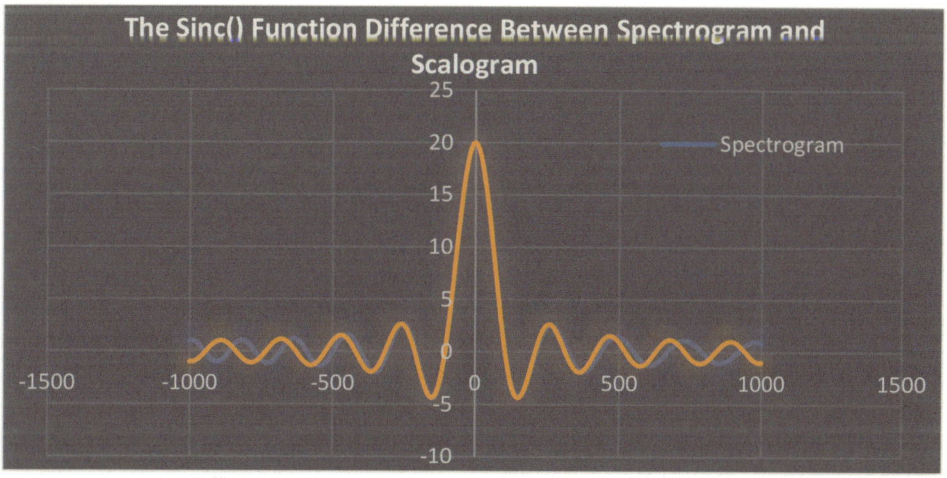

Figure 4.
The Sinc() *function for spectrogram and Scalogram.*

4. Connected graph approach for spectrogram and scalogram

Both spectrogram and scalogram are two-dimensional images and both have similar *Sinc()* function properties. Their image processing techniques are also similar. Therefore, only the spectrogram processing technique is introduced in this chapter to demonstrate the connected graph approach for FMRF signal processing. The scalogram processing is a straightforward extension.

4.1 Sparse cloud point representation of spectrograms

By binarizing a spectrogram, we can create a sparse cloud point representation (a binary image) of this spectrogram and call it the sparse time-frequency map. Thresholding and local maximum in the frequency direction can be used to create this sparse time-frequency map

$$M(\omega,t) = \begin{cases} I(\omega,t), & if\ I(\omega,t) > I(\omega-1,t), I(\omega-1,t)\ and\ T \\ 0, & otherwise \end{cases} \tag{19}$$

An FMRF signal, its spectrogram, and its sparse time-frequency map is shown in **Figure 5**. **Figure 5** shows that the nonzero points in the sparse time-frequency map created from the spectrogram of an FMRF signal form the time-frequency manifold that represents this FMRF signal. Since the nonzero pixels are a very small portion of the entire image of pixels and the connected graph approach, we are using only performs on these nonzero pixels, this connected graph approach has a very low computational cost.

4.2 The spectrogram and sparse time frequency map of a Noisy FMRF signal

We have discussed the spectrogram and sparse time-frequency map with no noises as shown in **Figure 5**. The spectrogram and its sparse time-frequency map for a noisy FMRF signal is shown in **Figure 6**.

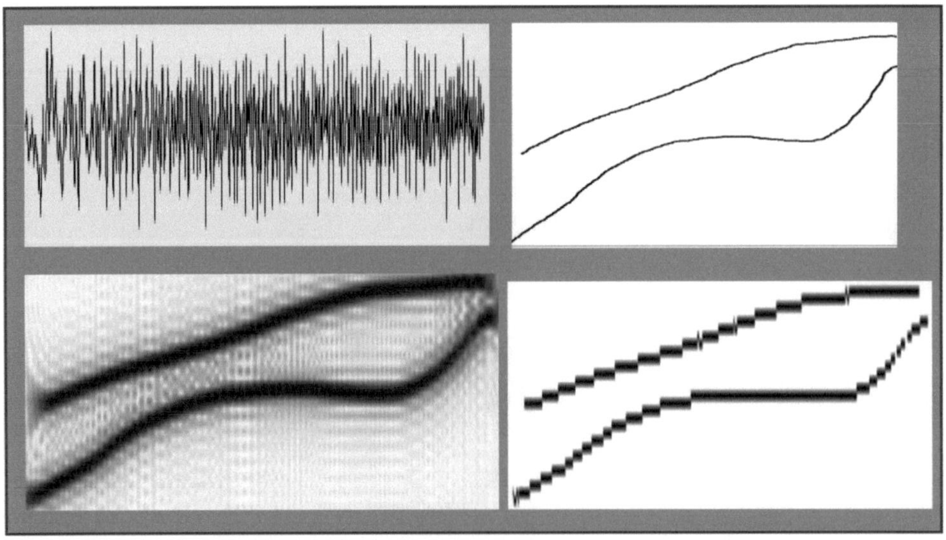

Figure 5.
An FMRF signal, and its spectrogram and sparse time-frequency map.

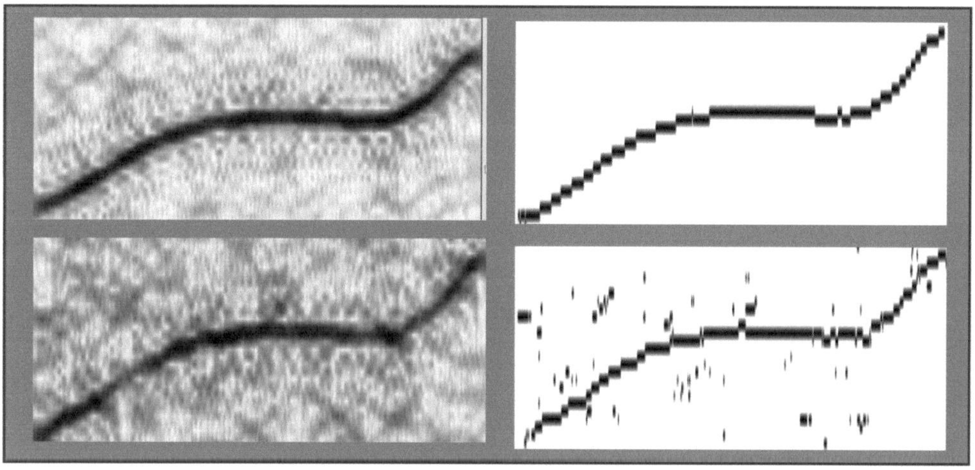

Figure 6.
Spectrogram and sparse time-frequency map of an FMRF signal with different noise levels: the top is for signal to noise ratio (SNR) = 6DB, and the bottom for SNR = 0 DB.

Figure 6 shows that the spectrograms and time-frequency maps for very noisy FMRF signals are similar to those without noises in **Figure 5**. The difference is that the time-frequency maps for noisy signals add some extra noise pixels. These noise pixels will be removed by the connected graph approach, however.

4.3 A graph approach for extracting time frequency manifolds

Figures 5 and **6** show that the sparse time-frequency map of an RF signal includes the points on the time-frequency manifold of this RF signal. A connected graph approach is used to extract this time-frequency manifold.

The graph to represent the sparse time-frequency map consists of nodes and edges. Each node n_i of the graph, corresponding to a nonzero pixel of the sparse time-frequency map, is represented by three variables. The first two variables x and y represent the pixel position for this node. The third member is a list of its neighbor nodes that are used to build connected graphs. A node n_i is defined with C++ as

```
struct Node{
        int x;
        int y;
        list <Node> nbrNodes; //the list of its neighbor nodes
};
```

Two nodes are connected if they are neighbors. For the node n_i, its neighbor nodes are found by checking the distance between two nodes. n_j is the neighbor of n_i if $(n_i.x - n_j.x) * (n_i.x - n_j.x) + (n_i.y - n_j.y) * (n_i.y - n_j.y) <$ Threshold Distance.

Each node is connected to its neighbors but disconnected to non-neighbor points. With this graph, the connected components can be found. Obviously, some connected graphs are the time-frequency manifolds as the FMRF signal, while others could be noises. Usually, small connected graphs are noises and can be removed.

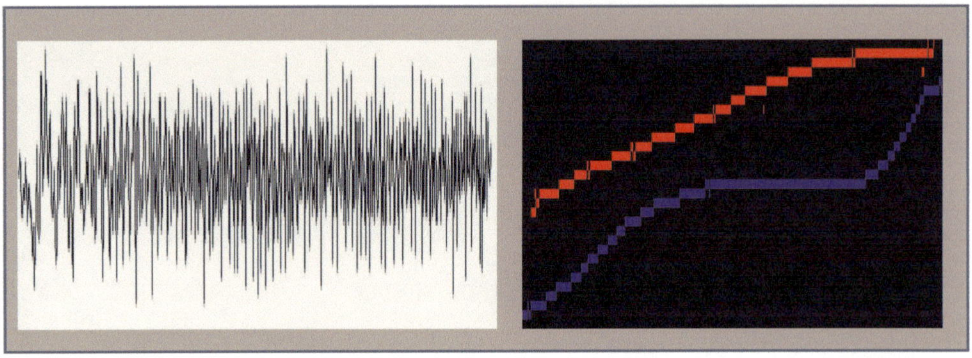

Figure 7.
*A two-source mixed FMRF signal and its time-frequency manifolds. These time-frequency manifolds are extracted by the connected graph approach from the sparse time-frequency map shown in the bottom right of **Figure** 5.*

The time-frequency manifolds for a two-source FMRF signal are extracted and shown in **Figure 7**, where two connected graphs are displayed for the time-frequency manifolds (red and blue) for two FMRF signals components.

Figure 7 shows that each individual component (red and blue) of the two-source cochannel and co-duration FMRF signals can be extracted using the connected graph approach.

4.4 Issues for the graph techniques

If the two components in a two-source FMRF signal are not connected to each other in their sparse time-frequency map, the connected graph approach is capable of extracting, separating, and classifying them, as is shown in **Figure** 7. However, when two or multiple components are connected to each other, as shown in **Figure 8**, the connected graph approach may not work well.

Figure 8 shows a two-source FMRF signal, its time-frequency manifold, spectrogram, and the time-frequency manifolds extracted by the graph approach.

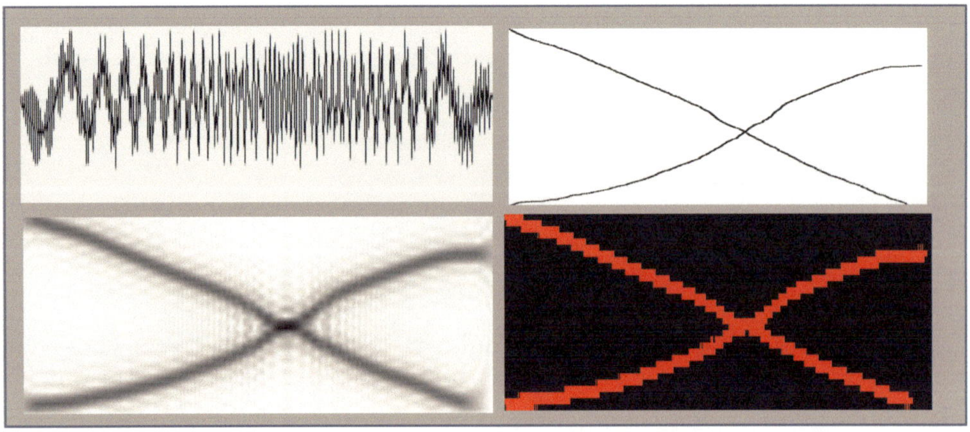

Figure 8.
A time-frequency manifold connected two source co-channel and co-duration FMRF signal, its time-frequency manifolds, spectrogram, and extracted manifolds.

One of these two source signals is a linear frequency modulation signal with a negative sweep rate (frequency decrease), and the other one is a nonlinear frequency modulation signal with a positive sweep rate (frequency increase). These two source FMRF signals are overlapped in both time and spectral space and form pulse-in-pulse signals. As is demonstrated in **Figure 8**, the connected graph approach cannot separate these two connected time-frequency manifolds. This inseparable problem causes serious issues for classifications and other RF signal processing. In the next section, a projection pursuit approach will be discussed to address this issue.

5. Projection pursuits approach for pulse-on-pulse FMRF signal processing

When the time-frequency manifolds of two FMRF components are crossed with each other, the spatial distance-based neighbor point definition has problems. These problems and their possible solutions are shown in **Figure 9**.

In **Figure 9**, the left figure defines the neighbor points in the graph approach by spatial distances. We call these neighbor points the spatial distance neighbor points. In this definition, the two manifolds are inseparable. Different from the spatial neighbor approach, a string neighbor point approach is used to build time-frequency manifolds. Two points are neighbors if these two points are spatial neighbors and if they are on the same string. The string neighbor approach is shown on the right side of **Figure 9**. **Figure 9** shows that the two manifolds are separable with the string neighbor approach even though they are inseparable from the spatial neighbor approach.

The projection pursuit approach is used to create string neighbor points. This approach is implemented in the following two steps:

Step 1. Create a graph for the time-frequency map by the spatial distance approach.

For each nonzero pixel, create a node n_i . For M nonzero pixels, $i = 0 \sim (M-1)$. The x, y member of a node n_i is the location of the pixel. The neighbor point set N_i of this node are generated by the spatial distance approach.

Step 2. Refine the neighbor points of each node n_i with the string approach.

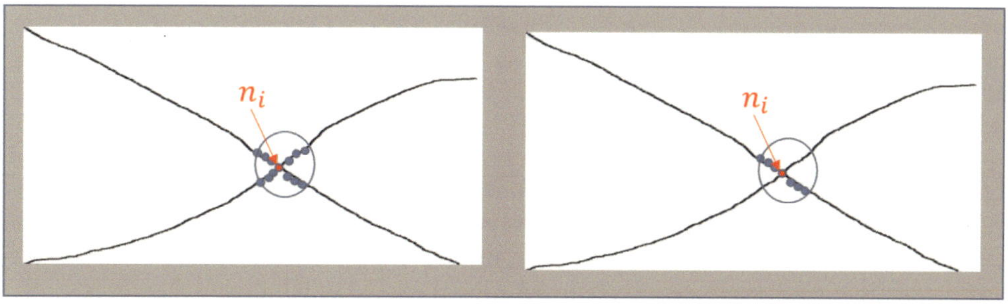

Figure 9.
Spatial neighbors and string neighbors. Spatial neighbors are defined by spatial distances and string neighbors are defined by both spatial distances and strings.

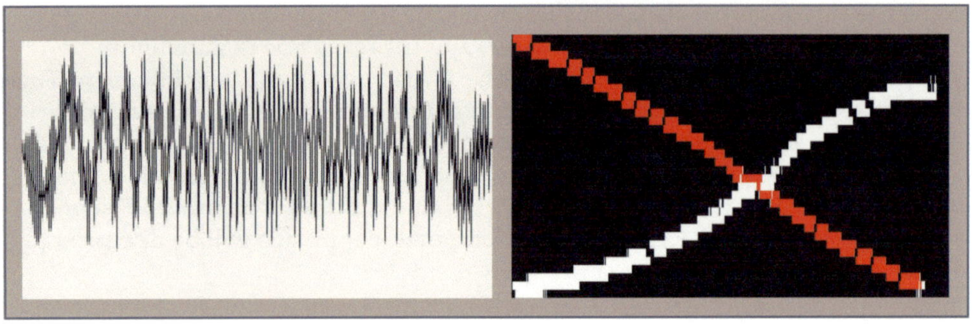

Figure 10.
A two-source mixed FMRF signal and its time-frequency manifolds extracted by graph and projection pursuits.

At the location of each node n_i, draw K lines $l_1, l_2, \ldots \ldots l_K$ with an increment angle $0, \frac{180}{K}, 2 * \frac{180}{K}, \ldots \ldots (K-1) * \frac{180}{K}$ to the horizon axis. Compute the projection to $l_1, l_2, \ldots \ldots l_K$ for each node of the neighbor nodes N_i. For each $l_i (i = 1 \sim K)$, add the top M ($<$ number of nodes in N_i) highest projections as the score S_i. Choose l_m so that its score S_m is largest. Then remove the neighbor nodes that have low projection values to l_m. After all the removing operations, the remaining neighbor nodes are the string neighbor nodes.

The above two steps are used to create string neighbor nodes. After the string neighbor nodes of the graph are created, the same connected graph approach discussed in Section 4 is used to create connected graphs and build the time-frequency manifolds for the FMRF signals.

The test results for the projection pursuits approach are shown in **Figure 10**. The right side of **Figure 10** shows two-time-frequency manifolds extracted by the projection pursuits approach. The red line is the down sweep linear frequency modulation component of this two-source FMRF signal while the white curve is the time-frequency manifold of the down sweep nonlinear frequency modulation component. It is shown from these test results that the projection pursuit approach is capable to separate and extract the time-frequency manifolds of complicated multisource FMRF signals (**Figure 10**).

6. Computational complexity analysis

Both spectrogram and scalogram approaches involve three components to perform their FMRF signal processing: transformation from an FMRF signal to a two-dimensional image, binarization of the image, and graph projection pursuit for creating the manifold of the FMRF signal.

Assume that the length of the signal to process is N. For the spectrogram approach, the transform from the FMRF signal to the spectrogram takes 0(NlogW) operations for a wind size W. The image size is W*(N/W) = N. Thus, the binarization takes 0(N) operations. Since the projection pursuits approach only processes a small fractional number of points in the image, its computational cost is much lower than 0(N). Putting the implementation of these three components together leads to the computational complexity 0(NlogW) for the spectrogram approach. Thus, the computation cost for the binarization and graph pursuits approach could be ignored when compared to the transform to create the spectrogram.

For the scalogram approach, since the transform from the FMRF signal to its scalogram image has a higher computational cost than the spectrogram approach and the same methods as the spectrogram approach are used for the binarization and graph projection pursuit, the computational complexity for the scalogram approach is the same as the computational complexity of the scalogram generation from the FMRF signal.

7. Conclusion

In this chapter, we introduce the spectrogram generation of an FMRF signal by using short-time Fourier transforms. Then, the spectrogram computation approach is extended to the scalogram computation by replacing the fixed size masks with frequency dependent masks.

Both spectrograms and scalograms are images, and a projection pursuits approach is introduced to process these images for separating and processing multisource cochannel and co-site FMRF signals.

It is shown that the projection pursuits method is very efficient, and its computational cost can be ignored when compared to the spectrogram or scalogram generation. Also, the projection pursuits approach is robust. It can separate and extract both non-connected and connected time-frequency manifolds for FMRF signal processing.

Author details

Bingcheng Li
Lockheed Martin, Owego, NY, USA

*Address all correspondence to: bing.li@lmco.com

References

[1] Mann S, Haykin S. Adaptive chirplet transform: An adaptive generalization of the wavelet transforms. Optical Engineering. 1992;**31**(6):1243-1256

[2] Mann S, Haykin S. The chirplet transform: A generalization of Gabor's logon transforms. In: Proc. Vision Interface. 1991. pp. 205-212

[3] Daudet SK, Gribonval R. Model-based matching pursuit – Estimation of chirp factors and scale of Gabor atoms with iterative extension. In: Proc. Signal process with Adapt Sparse Structured Representation; Rennes, France. 2005

[4] Pai A, Chassande-Mottin E, Rabaste O. Best network chirplet chain: Near-optimal coherent detection of unmodeled gravitational wave chirps with a network of detectors. Physical Review D. 2008;**77**(062005): 1-22

[5] Candes EJ, Charlton PR, Helgason H. Detecting highly oscillatory signals by chirplet path pursuit. Applied and Computational Harmonic Analysis. 2008;**24**(1):14-40

[6] Millioz F, Davies M. Sparse detection in the Chirplet transform: Application to FMCW radar signals. IEEE Transactions on Signal Processing. 2012;**60**(6): 2800-2813

[7] Mann S, Haykin S. The chirplet transform: Physical consideration. IEEE Transactions on Signal Processing. 1995; **43**(11):2745-2761

[8] Peng ZK, Meng G, Chu FL, Lang ZQ, Zhang WM, Yang Y. Polynomial chirplet transform with application to instantaneous frequency estimation. IEEE Transactions on Instrumentation and Measurement. 2011;**60**(9):3222-3229

[9] Yang Y, Zhang W, Peng Z, Meng G. Multicomponent signal analysis based on polynomial chirplet transform. IEEE Transactions on Industrial Electronics. 2012;**60**(9):3948-3956

[10] Tu X, Hu Y, Li F, Abbas S, Liu Y. Instantaneous frequency estimation for nonlinear FM signal based on modified polynomial Chirplet transform. IEEE Transactions on Instrumentation and Measurement. 2017;**66**(11):2898-2908

[11] Aoi M, Lepage K, Lim Y, Eden UT, Gardner TJ. An approach to time-frequency analysis with ridges of the continuous Chirplet transform. IEEE Transactions on Signal Processing. 2015; **63**(3):699-710

[12] Lim Y, Shinn-Cunningham B, Gardner T. Sparse contour representations of sound. IEEE Signal Processing Letters. 2012;**19**(10):684-687

[13] Li B. Polynomial chirplet approach for frequency modulation signal separation and classification. In: Proceedings Volume 11003, Radar Sensor Technology XXIII; 110031B. 2019

[14] Li B. Graph and projection pursuits approach for time frequency analysis. In: IEEE International Conference on Radar; Atlanta, GA; May 10-14, 2021.

[15] Li B. Time-frequency manifold representation for separating and classifying frequency modulation signals. In: Radar Sensor Technology XXV, SPIE Defense, Security, and Sensing; Orlando, FL. 2021

Section 2

Fundamentals of Spectral Analysis by Wavelet Transforms

Chapter 3

A Modern Review of Wavelet Transform in Its Spectral Analysis

Francisco Bulnes

Abstract

The spectral analysis, in much aspects as are the wavelet transform in its numerous versions and its relation with other transforms and special functions requires a special review, since the exploration in the frequency domain to the wavelet transform is more detailed and majorly more specific in different applications. For example, the wavelet transform of special function can be very useful to create and design special signal filters or, for example, to the interphase between reception-emission devices with sensorial parts of the human body. Also the quantum wavelet transform is very useful in the spectral study of traces of particles. Likewise, in this chapter, these aspects are considered as an inherent property of the wavelet transform in the spectral exploration of some phenomena. Finally, general results to the discrete case are given, which is analyzed to the wavelet transform and its spectra.

Keywords: discrete Fourier transform, discrete wavelet transform, fast Fourier transform, Gabor transform, short-time Fourier transform, spectra, quantum wavelet transform, wavelet transform

1. Introduction

Likewise, we consider a set of functions

$$\psi_{1k}, \psi_{2k}, \dots, \psi_{nk}, \dots \in L^2(R), \tag{1}$$

which define a Hilbert basis of square integrable functions [1]. Likewise, for each $j, k \in Z$, the ψ_{jk} represents dyadics and dilations of ψ, given by the functions:

$$\psi_{jk}(x) = 2^{\frac{j}{2}}\psi\left(2^j x - k\right), \tag{2}$$

$\forall j, k \in Z$. Likewise, for a function $\xi(x) \in L^2(R)$, and using the orthonormal functions family, by completeness we have:

$$\xi(x) = \sum_{-\infty}^{+\infty} c_{jk}\psi_{jk}(x), \tag{3}$$

Then the convergence of the series will be understood to be convergence in norm. Likewise, a representation of $\xi(x)$ is known as a wavelet series with wavelet coefficients c_{jk}. This implies that an orthonormal wavelet is self-dual. Then the wavelet integral transform is the integral transform [2] given for[1]:

$$W_{\psi}\xi = \frac{1}{\sqrt{|a|}}\int_{-\infty}^{+\infty}\bar{\psi}\left(\frac{x-b}{a}\right)\xi(x)\mathrm{dx}, \tag{4}$$

Here $a = 2^{-j}$ is the binary dilation or dyadic dilation, and $b = k2^{-j}$ is the binary or dyadic position. Then the wavelet transform can be modified depending on the response treatments that are given. For example in the images compression through impulse function $x(n) = \delta(n - n_i)$, for a discrete signal where impulse response can be used to evaluate the image compression-reconstruction system, the wavelet transform has been modified.

As has been said in different signal treatments, one of fundamental problems in electronics is obtaining a sufficiently clean signal in the different processes of communication, perception, and management of the signals in different continuous and discrete domains. For example, in signal processing in accelerometers for gait analysis, where actually is necessary to implement a good programming in real time for drones or other devices of vehicles, even human body parts with accelerations process in fault detections for design of low power pacemakers and also in ultrawideband in wireless [3]; the cleaning of signal is fundamental.

The wavelet transforms as transformation should allow only changes in time extension, but not shape. This could be affected by choosing suitable basis functions that allow for this. For example, in numerical analysis, we can consider the scale factor $c_n = c_0^n$, with the discrete frequency $\eta_m = \mathrm{mL}c_0^n$, having the wavelets (considering the discrete formula with the basis wavelet):

$$\Psi(k,n,m,) = \frac{1}{\sqrt{c_0^n}}\Psi\left[\frac{k - \mathrm{m}c_0^n}{c_0^n}L\right] = \frac{1}{\sqrt{c_0^n}}\Psi\left[\left(\frac{k}{c_0^n} - m\right)L\right], \tag{5}$$

where such discrete wavelets can be used through the discrete wavelet transform version:

$$W_D(n,m) = \frac{1}{\sqrt{c_0^n}}\sum_{k=0}^{K-1} w(k)\Psi\left[\left(\frac{k}{c_0^n} - m\right)L\right], \tag{6}$$

whose continuous (or analogic) is the standard wavelet transform:

$$W(c,\eta) = \frac{1}{\sqrt{c}}\int_{-\infty}^{+\infty} w(t)\Psi\left(\frac{t-\eta}{c}\right)\mathrm{dt}, \tag{7}$$

where c, is a scaling factor, and η, represents time shift factor. For the case (7) the Fourier transformation of signal $w(k)$, is computed with the FFT. An adequate selection of a discrete scaling factor $c_n$will be necessary. Changes in the time extension are expected to conform to the corresponding analysis frequency of the basis function, based on the uncertainty principle of signal processing.

---

[1] To recover the original signal $w(t)$, the first inverse continuous wavelet transform can be used: $\xi(\mathrm{t}) = \chi_{\psi}^{-1}\int_{-\infty}^{+\infty}\int_{-\infty}^{+\infty} \mathrm{W}_{\psi}\xi(\mathrm{a},\mathrm{b})\frac{1}{\sqrt{|\mathrm{a}|}}\psi\left(\frac{\mathrm{t}-\mathrm{b}}{\mathrm{a}}\right)\mathrm{db}\frac{\mathrm{da}}{\mathrm{a}^2}$.

For example, the wavelet transform of Shannon function can be very useful to creation of windows $\Psi^{\text{Sha}}(\omega)$, through functions $\text{Sha}(t)$, and with gate functions $\Pi(x)$, useful in the signal analysis by ideal band-pass filters that define a decomposition known as Shannon wavelets. Also, for example, the complex-valued Morlet wavelet is closely related to human perception, both hearing and the vision [4].

Likewise, the transition for "classical" wavelet transform (with some modifications accepted) to quantum wavelet transform can be approached by factoring the classical operators for the transformation into direct sums, direct products, and dot products of unitary matrices. Likewise, the permutation matrices play a vital role [5].

2. From the signal resolution problems until biological-sensorial perception

A fundamental property of the wavelet transform and the signal resolution problem can be discussed and explored simultaneously in time and frequency domains starting from the wavelet spectra:

$$W(\omega) = \frac{1}{\sqrt{|a|}} \int_{-\infty}^{+\infty} w\left(\frac{t-b}{a}\right) e^{-j\omega t} \mathrm{dt} = \sqrt{|a|} W(a\omega) e^{-j\omega t}, \tag{8}$$

where $W(\omega)$ is the Fourier transform of the basic wavelet $w(t)$. If the wavelets are normalized in terms of amplitude, the Fourier transforms of the wavelets with different scales will have the same amplitude that is suitable for implementation of the continuous wavelet transform using the frequency domain filtering. This property is fundamental in the samples of frequency pulses of signal spectra, since it shows that a dilatation t/a $(a > 1)$ of a function in the time domain produces a contraction $a\omega$, of its Fourier transform, which are "spectral wavelets" corresponding. Likewise, the term t/a has a metrology of frequency, which is equivalent to ω. However, in the technical convention, this term is known as scale, since the term "frequency" is reserved for the Fourier transform. Then the design of signal filters in frequency obey to the correlation between the signal and the wavelets, in the time domain that can be written as the inverse Fourier transform of the product of the conjugate Fourier transforms of the wavelets and the Fourier transform of the signal:

$$W_{\xi}(a,b) = \frac{\sqrt{|a|}}{2\pi} \int_{-\infty}^{+\infty} \Xi(\omega) W(j\omega a) e^{-j\omega t} d\omega \tag{9}$$

The Fourier transforms of the wavelets are referred to as the wavelet transform filters. The impulse response of the wavelet transform filter $\sqrt{|a|}W(a\omega)$ is the scaled wavelet $\frac{1}{\sqrt{|a|}} w\left(\frac{t}{a}\right)$. Therefore, the wavelet transform is a collection of wavelet transform filters with different scales, a. Then we can relate the short-time Fourier transform (STFT) [6] with the idea of the wavelets to determine the sinusoidal frequency and phase content of local sections of a signal considering as changes over time. If we introduce the Gaussian function, which can be regarded as a window function, then the STFT is the Gabor transform. Here the Gabor atoms or functions used to build from translations and modulations of generating function a family of functions are constructed and characterized.

Likewise, we can have direct applications of the STFT, to samples in real time of the complex processes, which require a speed compute of data through direct relation between machine and real-time domains in the measured and perception of the phenomena. Likewise, the STFT is performed on a computer using the fast Fourier transform (FFT), so both variables are discrete and quantized.

Secondly the Morlet transform is a Gabor transform consisting of a wavelet composed of a complex exponential (carrier) multiplied by a Gaussian window (envelope). This wavelet is closely related to human perception, both hearing [2] and vision. Then the functions related with these bio-sensorial perceptions use Sha(t) functions as special Gabor functions to discriminate steps of signal spectra in the perception and create of a signal response audible or visible required to the eye organ, the iris of eye, in the case of the vision and to audition, we have the audiphones that amplify the sounds to equilibrate the lack of the eardrum or other parts of middle and inner ear to perceive the sounds.

3. Some important results in discrete signal analysis

Let S^N be the complex sphere of dimension N, and let

$$\ldots, e^{-2\Omega j}, e^{-\Omega j}, 1, e^{\Omega j}, e^{-2\Omega j}, \ldots, \tag{10}$$

be a linear basis of signals space in $L^2[K]$ that generates the subspace W, such that $\forall w \in W$ is

$$w = \sum_{n=1}^{N} c_n e^{-n\Omega j}, \tag{11}$$

We define the space of nilpotent classes on $E_{[K]}$, (being $E_{[K]} = E_1 \oplus \ldots \oplus E_n$, [7] the total discrete signal space) with the component:

$$N\left(E_{[K]}\right) = \left\{F \in D'\left(G^0/K\right) | F = 0\right\}, \tag{12}$$

Proposition 3.1 [8, 9]. If $z \in N\left(E_{[K]}\right)$, and if $\beta \in H^i\left(n_0, L^2[K]\right)$, (then)

$$z\beta = e^{-n\Omega j} x[m] = p(z) x[n], \tag{13}$$

where $x[n]$ is a Gabor discrete function[2].

Proof. Here

$$n_0 \simeq n, \quad n_0 \simeq p_I,$$

[2] A discrete version of Gabor representation is

$x(t) = \sum_{n=-\infty}^{\infty} \sum_{m=\infty}^{\infty} C_{nm} g_{nm}(t)$,

with $g_{nm}(t) = s(t - m\eta_0)e^{-n\Omega jt}$.Similar to the DFT (discrete Fourier transformation) we have:

$x(k) = \sum_{n=-\infty}^{\infty} \sum_{m=\infty}^{\infty} C_{nm} g_{nm}(k)$,

where the Gabor basis functions are $g_{nm}(k) = s(k - m\eta_0)e^{-n\Omega jk}$.

where p_I is the unitary sphere $p \cap g_I$, where $g_I, = [g, g]$. We demonstrate on the dimension i, of the cohomological space $H^i(g_I, L^2[K])$. If $i = n = \dim n_0$, then

$$H^i(g_I, L^2[K]) = n_0 * \otimes L^2 / n_0 L^2, \tag{14}$$

Therefore, z acts by $I \otimes p(z)$. Then $p(z)$ acts for $(C \otimes \pi)p(z)$.(Then)

$$(I \otimes p(z))\beta = p_C(z)\beta, \tag{15}$$

As a special note, we have as a particular example an LTI system $L(e^{-n\Omega j}) = H(\Omega)e^{-n\Omega j}$, where $H(\Omega)$ is a projection of the system function.

Likewise, the result to $i = r + 1 \leq n$, then demonstrate for $i = r$. Let C be the periodic complex. Let $\alpha \in H * (X_i, Q_b Z_a)$, such that $\alpha(g \otimes v)gv$. Then α is the homeomorphism

$$\alpha : \mathrm{Hom}_K(p_I, C) \rightarrow \mathrm{Hom}_K(p_I, L^2[K]), \tag{16}$$

Let $X = \ker\alpha$, where specifically

$$X = \{\alpha \in \mathrm{Hom}_K(C, L^2[K]) | \alpha(g \otimes v) = 0, \quad \forall g \in U(g), v \in L^2[k]\}, \tag{17}$$

Then we have

$$0 \rightarrow X_i \rightarrow C \rightarrow L^2 \rightarrow 0, \tag{18}$$

Now $U(g)$, is a $U(p_I)$−complex free under left translations. Therefore

$$H^i(p_I, C) = 0, \tag{19}$$

$$\forall C = E_{[K]} \otimes Q_b Z_a, \text{(then)}$$

$$H^i(p_I, E_{[K]} \otimes Q_b Z_a) = 0, \tag{20}$$

$\forall j < n$, and $b \equiv a \bmod j$. Then the long exact sequence of discrete cohomology for this case of periodic complexes and $N(E_{[K]})$−complex is:

$$0 \rightarrow H^i(p_I, L^2[K]) \rightarrow H^{i+1}(p_I, X) \rightarrow H^{i+1}(p_I, E_{[K]} \otimes Q_b Z_a) \rightarrow 0, \tag{21}$$

where such injection implies the result.

A study realized in signal and systems analysis on a linear system can be approximated in the time-frequency domain due to the composition of an analysis filter-bank, a transfer matrix (sub-band model) and a synthesis filter-bank, which is a method known as sub-band technique.

In the varying case, time-frequency representations of LTV systems have connection with the Gabor expansion of signals through the corresponding integral equation. Then we will have an integral equation with Gabor function. For example, a work realized in that sense is the creation of 3D Gabor frame based in spatial spectral integral equation designed to solve the scattering from dielectric objects embedded in a multilayer medium. Likewise, this is based on the Gabor frame, as a new set of basic functions (belonging to a basis) [10] together with a set of equidistant Dirac-delta test functions.

Proposition 3.2. Exists an isomorphism given for the DFT that maps the proper nilpotent classes of the system controlled under transformations of p_I. Then DS-TFT is the FFT.

Let DFT be the isomorphism of the discrete signals:

$$E_{[K]} \rightarrow \tilde{E}_{[K]}, \tag{22}$$

where the explicit rule for any $\forall v \in L^2[K]$ is

$$\mathrm{DFT}(v) = (1/N)\mathrm{DFT}(v) = \mathrm{DWT}(v), \tag{23}$$

Then in each component of the space $E_{[K]}$, $(E_{[K]} = E_1 \oplus \ldots \oplus E_n.)$ we have:

$$F^{(k)} = 0, \tag{24}$$

where the DS-TFT satisfies in short-time interval. In each component, we have:

$$U(p_I)\mathrm{Hom}_K(F, L^2[K]) = \chi_\Lambda p_\xi, \tag{25}$$

which exists $\mathrm{FFT}^{\sim}(v)$, such that

$$\mathrm{DWT}(v) = \mathrm{FFT}(v), \tag{26}$$

More details of the demonstration can be consulted in [11].

4. Conclusions

In this introductory chapter, the various and several advantages of the wavelet transform and its properties on the signal and system analysis have been shown, considering different specialized window functions and the wavelet function basis. Likewise, wavelet analysis is known for its successful approach to solving the problem of signal analysis in both the time domain and frequency domain. Also, the analysis of the nonstationary signal generated by physical phenomena has a great challenge for various conversion techniques. In several studies, it has been shown that the transformation techniques such as Fourier transform and short Fourier transform fail to analyze nonstationary signals. But wavelet transform methods may be able to efficiently analyze both stable and unstable signals. All this the author develops with precision and accuracy. In the Gabor transform, the resolution analysis considers the uncertainty principles on nilpotent Lie groups and their corresponding algebras, which were established in the propositions given through spectral analysis given in the classes $H^i(p_I, L^2[K]), H^{i+1}(p_I, X)$, and $H^{i+1}(p_I, E_{[K]} \otimes Q_b Z_a)$. A scheme with neural network as components of a dynamical system can be proposed to demonstrate that using neural networks and linear filters in cascade and/or feedback configurations, a rich class of models of signaling and systematization in wide perspective and prospective can be constructed, considering the different filters designed by the different wavelet transform versions in short-time resolution or conventional resolution improving the canonical Fourier transform resolution. The multi-resolution analysis or multi-scale approximation can design a method considering practically the relevant discrete wavelet transforms (DWT), which can be considered as a fundamental set of special functions to realize approximations to solutions of different processes in time and the justification for the algorithm of the fast wavelet transform (FWT), for the calculating methods develop started with good wavelet bases.

Acknowledgements

I thank Eng. Rene Rivera-Roldán, Director of Electronics Engineering Program, for the support of hours for the investigation to carry out this work.

Nomenclature

STFT	Short-Time Fourier Transform
FFT	Fast Fouricr Transform
DFT	Discrete Fourier Transform
LTV	Linear-Time Varying System
DS − TFT	Discrete Short-Time Fourier Transform
$w(t)$	Basic wavelet
$W(\omega)$	Fourier transform of the basic wavelet $w(t)$
ψ_{jk}	Dyadic and dilations of the wavelets
$x[n]$	Discrete signal. In the proposition 2.1, is a discrete Gabor function
$E_{[K]}$	Discrete signal space. This space is a Hilbert space on the discrete domain K. Its orthogonal decomposing is $E_{[K]} = E_1 \oplus \dots \oplus E_n$. In the case of wavelets, the components $E_j (j = 1, 2, \dots, n)$ are dyadic translations and dilations of wavelet $w(k)$
DWT	Discrete Wavelet Transform

Author details

Francisco Bulnes
Research Department in Mathematics and Engineering, IINAMEI, TESCHA, Mexico

*Address all correspondence to: francisco.bulnes@tesch.edu.mx

References

[1] Reed M, Simon B. Methods of Modern Mathematical Physics. 1st ed. San Diego, California, USA: Academic Press Inc.; 1970

[2] Meyer Y. Wavelets and Operators. Cambridge, UK: Cambridge University Press; 1992

[3] Martin E. Novel method for stride length estimation with body area network accelerometers. 2011 IEEE Topical Conference on Biomedical Wireless Technologies, Networks, and Sensing Systems. 2011;**1**:79-82. DOI: 10.1109/BIOWIRELESS.2011.5724356

[4] Bruns A. Fourier-, Hilbert- and wavelet-based signal analysis: Are they really different approaches? Journal of Neuroscience Methods. 2004;**137**(2): 321-332

[5] Sharma J, Kumar A. Uncertainty Principles on Nilpotent Lie Groups, Journal of Representation Theory, arXiv: 1901.01676v1, [Math. R. T]. USA; 2019

[6] Allen JB. Short time spectral analysis, synthesis, and modification by discrete Fourier transform. IEEE Transactions on Acoustics, Speech, and Signal Processing. 1977;**ASSP-25**(3):235-238

[7] Akansu AN, Haddad RA. Multiresolution Signal Decomposition: Transforms, Subbands, and Wavelets. Boston, MA: Academic Press; 1992

[8] Bulnes F. Controlabilidad Digital Total sobre una Cohomología Discreta con Coeficientes en L2[K]. In: Proceedings of the Appliedmath III, International Conference in Applied Mathematics (APPLIEDMATH '03); 9–12 October 2007. Mexico City: IPN, UNAM, CINVESTAV, UNAM, Tec de Monterrey; 2007. pp. 71-77

[9] Bulnes F. Teoría de los (g, K)-Módulos. 1st ed. UNAM: Instituto de Matemáticas; 2000

[10] Dilz RJ, van Beurden MC. Fast operations for a Gabor-frame-based integral equation with equidistant sampling. IEEE Antennas and Wireless Propagation Letters. 2018;**17**(1):82-85 [8115279]

[11] Mallat S. A Wavelet Tour of Signal Processing. 2nd ed. San Diego, CA: Academic; 1999

Chapter 4

Bases of Wavelets and Multiresolution in Analysis on Wiener Space

Claude Martias

Abstract

The multiresolution analysis is applied into the space of square integrable Wiener functionals for extending well-known constructions of orthonormal wavelets in $L^2(\mathbf{R})$ to this space denoted by L^2 (μ), μ being the Wiener measure, as for instance Mallat's construction or furthermore Goodman–Lee and Tang construction. We also extend the Calderon–Zygmund decomposition theorem into the L^1(μ) framework. Even if L^1-spaces do not have unconditional bases, wavelets still outperform Fourier analysis in some sense. We illustrate this by introducing periodized Wiener wavelets.

Keywords: Wiener functionals, Wiener space, wavelet, multiresolution analysis

1. Introduction

The wavelet transform for Wiener functionals has been considered by the author and applied to diffusion processes and to the solutions to backward stochastic equation in [1]. This application is a purely mathematical one; others, having more practical aspect could be considered as an extension in networks domain (see [2]) for instance or in detection of change and chronological series analysis [3, 4]. Extension of the well-known concept in finite dimension of wavelet transform to analysis on Wiener space is one among many possible others, which could be useful to infinite dimensional analysis. The chapter devotes to this extension, more precisely, extension of multiresolution and bases of wavelets. We start by recalling notion of multiresolution analysis, exclusively on the space of square integrable Wiener functional. We follow the Mallat's construction [5] and notice that an extension to wavelets generated by a finite set of Wiener functionals can easily be done if we follow arguments of Goodman–Lee–Tang [6]. We give an example of multiresolution approximation generated by cardinal Hermite B-splines as in [6]. We complete our work by a study of unconditional bases for $L^p(\mu)$, $1<p<\infty$, μ being the Wiener measure. We start it by first proving an extension of the well-known Calderon–Zygmund decomposition theorem. As L^1-spaces do not have unconditional bases, we introduce a notion of "periodized Wiener wavelets" and show that wavelets still perform Fourier–Wiener analysis in some sense, as in finite dimension [7].

1.1 Multiresolution analysis in square Integrable Wiener Functionals

A multiresolution analysis in $L^2(\mu)$, the space of μ-square integrable Wiener functionals, μ denoting the Wiener measure, consists of a sequence approximation spaces $(\mathbf{V}_j)_{j\in\mathbb{Z}}$, $\mathbf{V}_j \subset L^2(\mu)$; these subspaces are assumed to be closed and satisfy the following:

$$\mathbf{V}_j \subset \mathbf{V}_{j-1}, \text{ for all } j \in \mathbb{Z}, \tag{1}$$

$$\text{the closed space generated by } \overline{\bigcup_{j\in\mathbb{Z}} \mathbf{V}_j} \text{ is } L^2(\mu), \tag{2}$$

$$\bigcap_{j\in\mathbb{Z}} \mathbf{V}_j = \{0\}. \tag{3}$$

If we denote by $\prod_j$ the orthogonal projection operator onto $\mathbf{V}_j$, then (2) ensures that $\lim_{j\to\infty}\prod_j \varphi = \varphi$ for all $\varphi \in L^2(\mu)$. There exist many ladders of spaces satisfying (1)–(3) that nothing to have with multiresolution; the multiresolution aspect is a consequence of the additional requirement

$$\varphi \in \mathbf{V}_j \Leftrightarrow \varphi(2^j.) \in \mathbf{V}_0. \tag{4}$$

That is, all the spaces are scaled versions of the central space $\mathbf{V}_0$. An example corresponding to the Haar multiresolution analysis in real analysis (see [8, 9]) is the following:

$$\mathbf{V}_j := \left\{\varphi \in L^2(\mu); \forall k \in \mathbb{Z} : \varphi_{\left[2^j e_{(k)}, 2^{j+1} e_{(k+1)}\right[} = constant\right\},$$

where $e(k), k \in \mathbb{Z}$ (or $e_k, k \in \mathbb{Z}$, another notation) is a fixed orthonormal basis in the Cameron – Martin space $\mathbf{H}$ defined by

$$\mathbf{H} := \left\{h : [0,1] \to \mathbb{R}^d / h(t) = \int_{[0,t]} h'(s)ds,\ \ (\|h\|_{\mathbf{H}})^2 := \int_{[0,1]} |h'(s)|^2 ds < \infty\right\},$$

and the above interval in definition of $\mathbf{V}_j$ is in the sense of usual order relation of functions. This example exhibits another feature that we require from a multiresolution analysis: invariance of $\mathbf{V}_0$ under " integer " translations,

$$\varphi \in \mathbf{V}_0 \Rightarrow \varphi(. - ne_0) \in \mathbf{V}_0, \forall n \in \mathbb{Z}. \tag{5}$$

Because of (4) this implies that if $\varphi \in \mathbf{V}_j$, then $\varphi(.-2^j ne_0)$ belongs to $\mathbf{V}_j$ for all $n \in \mathbb{Z}$. Finally, we require also that there exists $\phi \in \mathbf{V}_0$ so that

$$\{\phi_{0,n}; n \in \mathbb{Z}\} \text{ is an orthonormal basis in } \mathbf{V}_0, \tag{6}$$

where, for all $j, n \in \mathbb{Z}$, $\phi_{j,n}(\omega) := 2^{-j/2}\phi(2^{-j}\omega - ne_0)$. Together, (4) and (6) imply that $\left\{\phi_{j,n}; n \in \mathbb{Z}\right\}$ is an orthonormal basis for $\mathbf{V}_j$ for all $j \in \mathbb{Z}$. In the example given above, the possible choice for ϕ is the indicator Wiener functional for $[0, e_0] := \{\omega \in \mathbf{W}; 0 \le \omega(s) \le e_0(s),\ \ \forall s \in [0,1]\}$, that is, $\phi(\omega) = 1$ if $\omega \in [0, e_0]$, interval in the lattice space $(\mathbf{W},\le)$, and $\phi(\omega) = 0$ otherwise. We call ϕ the scaling Wiener functional of the multiresolution analysis. Note that ϕ depends on the choice of e_0. Hence, we will say that ϕ is the scaling Wiener functional in direction e_0.

The basic tenet of multiresolution analysis is that whenever a collection of closed subspaces satisfy (1)–(6), then there exists an orthonormal Wiener wavelet basis $\left\{\psi_{j,k}; j, k \in \mathbb{Z}\right\}$ of $L^2(\mu)$,

$\psi_{j,k}(\omega) := 2^{-j/2}\psi\left(2^{-j}\omega - ke_0\right)$, such that, for all $\varphi \in L^2(\mu)$,

$$\prod_{j-1}\varphi = \prod_{j}\varphi + \sum_{k\in\mathbb{Z}}\left\langle \varphi, \psi_{j,k}\right\rangle_{\mu}\psi_{j,k}, \tag{7}$$

where the bracket $\langle .,. \rangle_\mu$ denotes the scalar product in $L^2(\mu)$. For every $j \in \mathbb{Z}$, define $\mathbf{W}_j$ to be the complement orthogonal of $\mathbf{V}_j$ in $\mathbf{V}_{j-1}$. We have

$$\mathbf{V}_{j-1} = \mathbf{V}_j \oplus \mathbf{W}_j \tag{8}$$

and

$$\mathbf{W}_j \perp \mathbf{W}_{j'} \text{ if } j \neq j'. \tag{9}$$

It follows that, for $j < j_0$,

$$\mathbf{V}_j = \mathbf{V}_{j0} \oplus \left(\oplus_{k=0}^{j(0)-1}\mathbf{W}_{j(0)-k}\right), \tag{10}$$

where all these subspaces are orthogonal. By virtue of (2) and (3), this implies

$$L^2(\mu) = \bigoplus_{k\in\mathbb{Z}} \mathbf{W}_j, \tag{11}$$

A decomposition of $L^2(\mu)$ into mutually orthogonal subspaces. Furthermore, the $\mathbf{W}_j$ spaces inherit the scaling property (4) from the $\mathbf{V}_j$:

$$\varphi \in \mathbf{W}_j \Leftrightarrow \varphi(2^j.) \in \mathbf{W}_0. \tag{12}$$

Formula (7) is equivalent to saying that, for fixed j, $\left\{\psi_{j,k}; k \in \mathbb{Z}\right\}$ constitutes an orthonormal basis for $\mathbf{W}_j$. Because of (11), (2) and (3), this then automatically implies that the whole collection $\left\{\psi_{j,k}; j, k \in \mathbb{Z}\right\}$ is an orthonormal basis for $L^2(\mu)$. On the other hand, (12) ensures that if $\{\psi_{0,k}; k \in \mathbb{Z}\}$ is an orthonormal basis for $\mathbf{W}_0$, then $\left\{\psi_{j,k}; k \in \mathbb{Z}\right\}$ will likewise be an orthonormal basis for $\mathbf{W}_j$, for any $j \in \mathbb{Z}$. Construction of ψ can be done as in the case of real analysis, using Fourier–Wiener transform (see [10] for an introduction to this notion) in the place of Fourier transform [11].

Our task thus reduces to finding $\psi \in \mathbf{W}_0$ such that the $\psi(.-ke_0)$ constitute an orthonormal basis for $\mathbf{W}_0$. Let us first write out some interesting properties of ϕ and $\mathbf{W}_0$.

1. Since $\phi \in \mathbf{V}_0 \subset \mathbf{V}_{-1}$, and the $\phi_{-1,n}$ are an orthonormal basis in $\mathbf{V}_{-1}$, we have

$$\phi = \sum_n c_n \phi_{-1,n}, \tag{13}$$

with

$$c_n = \langle \phi, \ \phi_{-1,n} \rangle_\mu \text{ and} \sum_n |c_n|^2 = 1. \tag{14}$$

we can rewrite (13) as either

$$\phi(\omega) = 2^{1/2} \sum_n c_n \phi(2\omega - ne_0) \tag{15}$$

or

$$\hat{\phi}(\xi) = 2^{-1/2} \sum_n \exp\left[-\mathrm{i}n\langle \xi, e_0 \rangle_{\mathbf{H}}/2\right] \hat{\phi}(\xi/2), \tag{16}$$

where the convergence in either sum holds in $L^2(\mu)$—sense, $\hat{\phi}$ denoting the Fourier–Wiener transform. Formula (16) can be rewritten as

$$\hat{\phi}(\xi) = m_0(\xi/2)\hat{\phi}(\xi/2), \tag{17}$$

where

$$m_0(\xi) = 2^{-1/2} \sum_n c_n \exp\left[-\mathrm{i}n\langle \xi, e_0 \rangle_{\mathbf{H}}\right]. \tag{18}$$

Equality in (17) holds pointwise μ—almost everywhere.

2. The orthonormality of the $\phi(.-ke_0)$ leads to special properties for m_0. We have

$$\begin{aligned}
\delta_{k,0} &= \int_{\mathbf{H}} \left|\hat{\phi}(\xi)\right|^2 \exp\left[\mathrm{i}k\langle \xi, e_0 \rangle_{\mathbf{H}}\right] \mu(d\xi) \\
&= \sum_l \int_{\{\langle \xi, e(0)\rangle \in [2\pi l, 2\pi(l+1)[\}} \left|\hat{\phi}(\xi)\right|^2 \exp\left[\mathrm{i}k\langle \xi, e_0 \rangle_{\mathbf{H}}\right] \mu(d\xi) \\
&= \sum_l \exp\left[-2\pi l^2\right] \int_{\{\langle h, e(0)\rangle \in [0, 2\pi[\}} \left|\hat{\phi}(h + 2\pi l e_0)\right|^2 \exp\left[\mathrm{i}k\langle h, e_0 \rangle_{\mathbf{H}}\right]. \\
&\quad \exp\left[-2\pi l \int_{[0,1[} \underset{0}{e}(s) dh(s)\right] \mu(dh) \\
&= \int_{\{\langle h, e(0)\rangle \in [0, 2\pi[\}} \sum_l \left|\hat{\phi}(h + 2\pi l e_0)\right|^2 \exp\left[\mathrm{i}k\langle h, e_0 \rangle_{\mathbf{H}}\right] \mu(dh),
\end{aligned}$$

Implying

$$\sum_l \left|\hat{\phi}(h + 2\pi l e_0)\right|^2 = (2\pi)^{-1} \mu - \text{a.e.}. \tag{19}$$

Substituting (17) leads to

$$\sum_l |m_0(\xi + \pi l e_0)|^2 \ \left|\hat{\phi}(\xi + \pi l e_0)\right|^2 = (2\pi)^{-1};$$

Splitting the sum into even and odd l, using the periodicity of m_0 and applying (19) gives

$$|m_0(\xi)|^2 + |m_0(\xi + \pi e_0)|^2 = 1\,\mu - \text{a.e.}. \tag{20}$$

1. Let us now characterize $\mathbf{W}_0$. $\varphi \in \mathbf{W}_0$ is equivalent to $\varphi \in \mathbf{V}_{-1}$ and φ is orthogonal to $\mathbf{V}_0$. Since $\varphi \in \mathbf{V}_{-1}$, we have
$\varphi = \sum_n \lambda_n \Phi_{-1,n}$, with $\lambda_n = \langle \varphi, \ \phi_{-1,n} \rangle_\mu$. This implies

$$\hat{\varphi}(\xi) = 2^{-1/2} \sum_n \lambda_n \exp\left(-\text{in}\langle \xi, e_0 \rangle_{\mathbf{H}}/2\right) \hat{\Phi}(2^{-1}\xi) = m_\varphi(2^{-1}\xi)\hat{\Phi}(2^{-1}\xi), \tag{21}$$

where

$$m_\varphi(\xi) = 2^{-1} \sum_n \lambda_n \exp\left(-\text{in}\langle \xi, e_0 \rangle_{\mathbf{H}}\right); \tag{22}$$

m_φ is a $2\pi e_0$-periodic Wiener functional; convergence in (22) holds pointwise μ—a. e.. The constraint " φ orthogonal to $\mathbf{V}_0$ " implies that φ is orthogonal to $\Phi_{0,k}$ for all k, that is,

$$\int_{\mathbf{H}} (\hat{\varphi}(h)) \left(\overline{\hat{\Phi}(h)}\right) \ \exp\left(ik\langle h, e_0 \rangle_{\mathbf{H}}\right) \mu(dh) = 0$$

where z^c denotes the conjugate complex of complex z, or

$$\int_{\{\langle h, e_0 \rangle \in [0, 2\pi[\}} \exp\left(\text{ik}\langle h, e_0 \rangle_{\mathbf{H}}\right) \sum_l \left(\hat{\varphi}(h + 2\pi l e_0)\overline{\hat{\Phi}(h + 2\pi l e_0)}\right) \mu(dh) = 0;$$

note that, out of the ordinary, we denoted by $h \in \mathbf{H}$ the integration variable. Hence

$$\sum_l \left(\hat{\varphi}(h + 2\pi l e_0)\overline{\hat{\phi}(h + 2\pi l e_0)}\right) = 0, \tag{23}$$

where the series in (23) converges absolutely in $L^1(\mu)$. Substituting (17) and (21), regrouping the sums for odd and even l (which we are allowed to do as we have an absolutely convergence), and using (19) leads to

$$m_\varphi(h)\left(\overline{m_0(h)}\right) + m_\varphi(h + \pi e_0)\left(\overline{m_0(h + \pi e_0)}\right) = 0\,\mu - \text{a.e.}. \tag{24}$$

Since $\left(\overline{m_0(h)}\right)$ and $\left(\overline{m_0(h + \pi e_0)}\right)$ cannot vanish together on a set of nonzero Wiener measure (because of (20)), this implies the existence of a $2\pi e_0$-periodic Wiener functional Θ so that

$$m_\varphi(h) = \Theta(h)\left(\overline{m_0(h + \pi e_0)}\right)\,\mu - \text{a.e.} \tag{25}$$

and

$$\Theta(h) + \Theta(h + \pi e_0) = 0\ \mu - \text{a.e.}. \tag{26}$$

This last equation can be rewrite as

$$\Theta(h) = e\big(i\langle h, e_0\rangle_{\mathbf{H}}\big)\nu(2h), \tag{27}$$

where ν is a $2\pi e_0$-periodic Wiener functional.
Substituting (25, 27) into (21) gives

$$\hat{\varphi}(\xi) = \exp\left(\mathrm{i}\langle h, e_0\rangle_{\mathbf{H}}/2\right)\left(\overline{m_0(\xi/2 + \pi e_0)}\right)\nu(\xi)\hat{\Phi}(\xi/2). \tag{28}$$

3. The general form (28) for the Fourier–Wiener transform of $\varphi \in \mathbf{W}_0$ suggests that we take

$$\hat{\Psi}(\xi) = \exp\left(\mathrm{i}\langle h, e_0\rangle_{\mathbf{H}}/2\right)\left(\overline{m_0(\xi/2 + \pi e_0)}\right)\hat{\Phi}(\xi/2) \tag{29}$$

as a candidate for our wavelet. Forgetting convergence questions, (28) can indeed be written as

$$\hat{\varphi}(\xi) = \left(\sum_k \nu_k \exp\left(-\mathrm{ik}\langle \xi, e_0\rangle_{\mathbf{H}}\right)\right)\hat{\Psi}(\xi)$$

or

$$\varphi = \sum_k \nu_k \Psi(.\!-\!ke_0),$$

so that the $\Psi(.\!-\!ne_0)$ are a good candidate for a basis of $\mathbf{W}_0$. We need to verify that the $\Psi_{0,k}$ are indeed an orthonormal basis for $\mathbf{W}_0$. First, the properties of m_0 and $\hat{\Phi}$ ensure that (29) defines an $L^2(\mu)$-Wiener functional belonging to $\mathbf{V}_{-1}$ and orthogonal to $\mathbf{V}_0$ (by the analysis above), so that $\Psi \in \mathbf{W}_0$. Orthonormality of the $\Psi_{0,k}$ is easy to check:

$$\int_{\mathbf{H}} \Psi(h)\left(\overline{\Psi(h\!-\!me_0)}\right)\mu(dh) = \int_{\mathbf{H}} \left|\hat{\Psi}(\xi)\right|^2 \exp\left(\mathrm{im}\langle \xi, e_0\rangle_{\mathbf{H}}\right)\mu(d\xi)$$
$$= \int_{\{\langle \xi, e_0\rangle \in [\mathbf{0}, \mathbf{2\pi}[\}} \exp\left(\mathrm{im}\langle \xi, e_0\rangle_{\mathbf{H}}\right)\sum_l \left|\hat{\Psi}(\xi + 2\pi l e_0)\right|^2 \mu(d\xi).$$

Now

$$\sum_l \left|\hat{\Psi}(\xi + 2\pi l e_0)\right|^2 = \sum_l |m_0(\xi/2 + \pi l e_0 + \pi e_0)|^2 \left|\hat{\Phi}(\xi/2 + \pi l e_0)\right|^2$$
$$= |m_0(\xi/2 + \pi e_0)|^2 \sum_n \left|\hat{\Phi}(\xi/2 + 2\pi n e_0)\right|^2$$
$$+ |m_0(\xi/2)|^2 \sum_n \left|\hat{\Phi}(\xi/2 + \pi e_0 + 2\pi n e_0)\right|^2$$
$$= (2\pi)^{-1}\left(|m_0(\xi/2)|^2 + |m_0(\xi/2 + \pi e_0)|^2\right)\mu - \text{a.e.}\,(\text{by } (1.19))$$
$$= (2\pi)^{-1}\mu - \text{a.e.}\,(\text{by } (1.20)).$$

Hence $\int_{\mathbf{H}} \Psi(h)\left(\overline{\Psi(h-me_0)}\right)\mu(dh) = \delta_{m,0}$. In order to check that the $\Psi_{0,m}$ are indeed a basis for all $\mathbf{W}_0$, it then suffices to check that any $\varphi \in \mathbf{W}_0$ can be written as $\varphi = \sum_n \gamma_n \Psi_{0,n}$, with $\sum_n |\gamma_n|^2 < \infty$, or

$$\hat{\varphi}(\xi) = \gamma(\xi)\hat{\Psi}(\xi), \tag{30}$$

with γ a square integrable $2\pi e_0$-periodic Wiener functional. Let us return to (28). We have $\hat{\varphi}(\xi) = \nu(\xi)\hat{\Psi}(\xi)$, with $\nu \in L^2(\mu)$. By (22),

$$\int_{\mathbf{H}} |m_\varphi(\xi)|^2 \mu(d\xi) = \pi \sum_n |\lambda_n|^2 = \pi\left(\|\varphi\|_\mu\right)^2 < \infty.$$

On the other hand, by (25),

$$\int_{\mathbf{H}} |m_\varphi(\xi)|^2 \mu(d\xi) = \int_{\mathbf{H}} |\Theta(\xi)|^2 |m_0(\xi + \pi e_0)|^2 \mu(d\xi)$$

$$= \int_{\mathbf{H}} |\Theta(\xi + \pi e_0)|^2 |m_0(\xi + \pi e_0)|^2 \mu(d\xi), \text{ using } (1.26) \tag{31}$$

Now, with the change of variable $h = \xi + \pi e_0$ and with the help of stochastic calculus we find for this integral

$$\exp\left[\left(1-\pi^2\right)/2\right] \int_{\mathbf{H}} |\Theta(h)|^2 |m_0(h)|^2 \mu(dh)$$

$$= \exp\left[\left(1-\pi^2\right)/2\right] \int_{\mathbf{H}} |\Theta(h)|^2 \left(1 - |m_0(h + \pi e_0)|^2\right)\mu(dh), \text{ using } (1.20),$$

$$= \exp\left[\left(1-\pi^2\right)/2\right] \int_{\mathbf{H}} |\Theta(h)|^2 \mu(dh)$$

$$-\exp\left[\left(1-\pi^2\right)/2\right] \int_{\mathbf{H}} |\Theta(h + \pi e_0)|^2 |m_0(h + \pi e_0)|^2 \mu(dh).$$

Put: $I := \int_{\mathbf{H}} |\Theta(\xi + \pi e_0)|^2 |m_0(\xi + \pi e_0)|^2 \mu(d\xi)$. We hence have, combining (31) and this last equality:

$I = \exp[(1-\pi^2)/2] \int_{\mathbf{H}} |\Theta(h)|^2 \mu(dh) - \exp[(1-\pi^2)/2] I$, which implies

$$I = \exp\left[\left(1-\pi^2\right)/2\right].(1 + \exp\left[\left(1-\pi^2\right)/2\right]^{-1} \int_{\mathbf{H}} |\Theta(h)|^2 \mu(dh).$$

Hence, $\left(\|\nu\|_\mu\right)^2 = 2\pi\left(\|\varphi\|_\mu\right)^2 < \infty$, and φ is of the form (30) with $\gamma \in L^2(\mu)$ and $2\pi e_0$-periodic. We have thus prove the following theorem.

Theorem 1.1. If a ladder of closed subspaces $(\mathbf{V}_j)_{j\in\mathbb{Z}}$ in $L^2(\mu)$ satisfies the conditions (1)–(6), then there exists an orthonormal Wiener wavelet basis $\{\Psi_{j,k}; j,k\in\mathbb{Z}\}$ for $L^2(\mu)$ such that

$$\Pi_{j-1} = \Pi_j + \sum_k \langle ., \Psi_{j,k}\rangle_\mu \Psi_{j,k}. \tag{32}$$

One possibility for the construction of the Wiener wavelet Ψ is,

$$\hat{\Psi}(\xi) = \exp\left(i\langle\xi, e_0\rangle_{\mathbf{H}}/2\right)\left(\overline{m_0(\xi/2+\pi e_0)}\right)\hat{\Phi}(\xi/2),$$

(with m_0 as defined by (14) and (18)), or equivalently

$$\Psi = \sum_n (-1)^n c_{-n-1}\Phi_{-1,n}, \tag{33}$$

$\Psi(\omega) = 2^{-1/2}\sum_n (-1)^n c_{-n-1}\ \Phi(2\omega{-}ne_0)$, with convergence in this series in $L^2(\mu)$ – sense.

All the argument we hold for the proof of the above theorem is exactly the same which can be found in any book on this subject (in the finite dimension case). We can hence follow the Mallat's construction [5], *via* a multiresolution analysis, of orthonormal wavelets for μ-square integrable Wiener functionals. The reader can also take a look on Meyer's books [12, 13]. An extension to wavelets generated by a finite set of Wiener functionals can easily be done following arguments of Goodman–Lee and Tang paper [6]. We give now in next section an example of multiresolution approximation generated by cardinal Hermite *B*-splines in $L^2(\mu)$, as we can find it in [6] for the one-dimensional case.

1.2 Multiresolution approximation generated by cardinal B-splines in $L^2(\mu)$

Let us first beginning with some recalls. For n, r positive integers, n even, such that $n \geq 2r$, put

$$\underset{n}{S} := \left\{f\in C^{n-r-1}(\mathbb{R}) : f_{|[v,v+1[} \in \mathbf{P}_{n-1}, v\in\mathbb{Z}\right\}, \tag{34}$$

where $\mathbf{P}_{n-1}$ is the class of polynomials of degree $\leq n{-}1$. Functions in S^r_n are called cardinal Hermite splines of degree≤ than n–1.

For $j = 0, \ldots, r{-}1$, let

$$S^r_{n,j} := \left\{f\in S^r_n : f^{(k)}(v) = 0, v\in\mathbb{Z}, k = 0, \ldots, r{-}1, k\neq j\right\}. \tag{35}$$

The space $S^r_{n,j}$ has a basis consisting of integer translates of a function $N_{n,j} = N^r_{n,j}\in S^r_{n,j}, j = 0, \ldots, r{-}1$, with minimal support $[-n/2{-}1+r, n/2+1-r]$, in the sense that every $f\in S^r_{n,j}$ has a unique representation of the form

$$f(x) = \sum_{v\in\mathbb{Z}} c_v N_{n,j}(x{-}v), x\in\mathbb{R}, \tag{36}$$

(see [14]). The functions $N_{n,j}$ are called cardinal Hermite B-splines and their Fourier transforms are given by (see [14])

$$\hat{N}_{n,j}(u) = [2\sin(u/2)]^n |H_{r,j}(\alpha_n(u))| \tag{37}$$

where $H_{r,j}(\alpha_n(u))$ denotes the matrix obtained from the Hankel matrix $H_r(\alpha_n(u))$ of order r by replacing the $(j+1)$ th column by $(u^n,\ u^{n-1}, \dots, u^{n-r+1})^T$, $j = 0, \dots, r{-}1$, denoting by $|H_r(\alpha_n(u))|$ its determinant.

Consider the map $K : \ell^2(\mathbb{Z})^r \to L^2(\mu)$ defined by the following:

$$K(s)(\omega(.)) := \sum_{j:0\to r-1} \sum_{v\in\mathbb{Z}} s_j(v)\mathbf{T}^v N_{n,j}(\omega(.)), \omega\in C([0,1]), \tag{38}$$

$\mathbf{T}$ being a unitary operator on $L^2(\mu)$.

Theorem 2.1. The map K defined by (38) is an isomorphism of $\ell^2(\mathbb{Z})^r$ onto a subspace of $L^2(\mu)$.

This above theorem is an easy consequence of Theorem 4.1 in [6]. Let us denote by $\tilde{S_n^r}(\mu)$ the range of K. This is a closed subspace of $L^2(\mu)$. Furthermore, $\tilde{S_n^r}(\mu) = S_n^r(\mu) \cap L^2(\mu)$ where $S_n^r(\mu)$ is the space deduced from S_n^r by the following: $\varphi \in S_n^r(\mu) \Leftrightarrow \varphi(\omega(t)) \in S_n^r$, $\omega \in C([0,1])$, $t \in [0,1]$. Therefore, if $D\varphi(\omega) := \varphi(2\omega)$, we have as in real analysis:

The closed space generated by $\overline{\bigcup_{m\in\mathbb{Z}} D^m \tilde{S_n^r}(\mu)}$ in $L^2(\mu)$ contains the one generated by $\overline{\bigcup_{m\in\mathbb{Z}} D^m \tilde{S_n^1}(\mu)}$ which is $L^2(\mu)$.

We also have:

$$\bigcap_{m\in\mathbb{Z}} D^m \tilde{S_n^r}(\mu) = \{0\}.$$

Let $\mathbf{V}_m := D^m\ \tilde{S_n^r}(\mu)$, $m \in \mathbb{Z}$. Then, $(\mathbf{V}_m)_{m\in\mathbb{Z}}$ is a multiresolution approximation of $L^2(\mu)$. We shall call $(\mathbf{V}_m)_{m\in\mathbb{Z}}$ a Wiener Hermite spline multiresolution approximation of $L^2(\mu)$. We could go on and hence extend all results of T.N.T. Goodman, S.L. Lee and W.S. Tang paper [6]. We now prefer in next section deal with unconditional bases of $L^p(\mu)$.

1.3 Unconditional bases for $L^p(\mu)$

Orthonormal bases of wavelets in $L^2(\mu)$ give good (i.e., unconditional) bases for many other spaces than L^2. We start by proving the following extension of Calderon–Zygmund decomposition theorem. We denote in this section by $|\omega|$ for $\omega \in W := C_0([0,1]; \mathbb{R}^d)$ the sup norm [15, 16].

Theorem 3.1. Assume φ to be a positive Wiener functional in $L^1(\mu)$.
Fix $\alpha > 0$. Then the Wiener space $\mathbf{W} := C_0([0,1]; \mathbb{R}^d)$ can be decomposed as follows:

1. $\mathbf{W} = \mathbf{G} \cup \mathbf{B}$ with $\mathbf{G} \cap \mathbf{B} = \emptyset$;

2. On the " good " set $\mathbf{G}$, $\varphi(\omega) \le \alpha\ \mu$ - a. e.;

3. The " bad " set **B** can be written as $\mathbf{B} = \underset{k \in \mathbb{N}}{\cup} \mathbf{Q}_k$, where the $\mathbf{Q}_k$ are non-overlapping intervals in the Banach lattice **W**, and

$$\alpha \leq \mu(\mathbf{Q}_k)^{-1} \int_{\mathbf{Q}_k} \varphi(\omega)\, \mu(d\omega) \leq 2\alpha, \forall k \in \mathbb{N}.$$

The proof of this theorem is identical with the proof of Theorem 9.1.1, p.289 of chapter 9 in Daubechies book [8, 17]. Next, we define Calderon–Zygmund operators for square integrable Wiener functionals and extend a classical property [18].

Definition3.2. A Calderon–Zygmund operator T on the Wiener space **W** is an integral operator

$$(T\varphi)(\omega) = \int_{\mathbf{H}} K(\omega, \xi)\mu(d\xi) \tag{39}$$

for which the integral kernels satisfies

$$|K(\omega, \xi)| \leq C/|\omega - \xi|, \tag{40}$$

$$\|\nabla_\omega K(\omega, \xi)\|_{\mathbf{H}} + \|\nabla_\xi K(\omega, \xi)\|_{\mathbf{H}} \leq C/|\omega - \xi|, \tag{41}$$

where the derivation symbol ∇ is the Malliavin derivative, and which defines a bounded operator on $L^2(\mu)$.

Theorem 3.3. A Calderon–Zygmund operator on **W** is also a bounded operator from $L^1(\mu)$ to $L^1_{weak}(\mu)$.

We recall below the definition of $L^1_{weak}(\mu)$.

Definition 3.4. $\varphi \in L^1_{weak}(\mu)$ if there exists $C > 0$ so that, for all $\alpha > 0$,

$$\mu\{\omega \in \mathbf{W}/|\varphi(\omega)| \geq \alpha\} \leq C/\alpha. \tag{42}$$

Like the proof of Theorem 3.1, this theorem is the extension of Theorem 9.1.2, p.291, chapter 9 in Daubechies book [8, 17]. We do not reproduce it as it is identical to the proof of Theorem 9.1.2 in [8]. The infimum of all C for which (42) holds (for all $\alpha > 0$) will be called $\|\varphi\|_{L^1_{weak}}$. Note that this notation is abusive as it is not a " true " norm.

Now, let T be a Calderon–Zygmund operator on **W**. As T maps $L^2(\mu)$ to $L^2(\mu)$ and $L^1(\mu)$ to $L^1_{weak}(\mu)$, we can extend T to other $L^p(\mu)$-spaces by interpolation theorem of Marcinkiewicz.

Theorem 3.5. If an operator T satisfies

$$\|T\varphi\|_{Lq(1)(\mu)weak} \leq C_1 \|\varphi\|_{Lp(1)(\mu)}, \tag{43}$$

$$\|\varphi\|_{Lq(2)(\mu)weak} \leq C_2 \|\varphi\|_{Lp(2)(\mu)}, \tag{44}$$

where $q_1 \leq p_1$, $q_2 \leq p_2$, then for $1/p = t/p_1 + (1-t)/p_2$,

$1/q = t/q_1 + (1-t)/q_2$, with $0 < t < 1$, there exists a constant K, depending on p_1,q_1, p_2,q_2, and t, so that

$$\|T\varphi\|_{Lq(\mu)} \leq K \|\varphi\|_{Lp(\mu)}.$$

Here $L_q(\mu)_{weak}$ stands for the space of all Wiener functionals φ for which $\|\varphi\|_{Lq(\mu)weak} := (\inf \{C/\mu\{\omega; |\varphi(\omega)| \geq \alpha\} \leq C\alpha^{-q} \text{ for all } \alpha > 0\})^{1/q}$ is finite.

The proof of this theorem can be found in E. Stein and G. Weiss [19] for the finite dimensional case. Extension to Wiener functionals can easily be done from this proof. Notice that with this theorem it only needs weaker bounds at the two extrema, and nevertheless derives bounds on $L^q(\mu)$-norms (not $L^q_{weak}(\mu)$) for intermediate values q. The Marcinkiewicz interpolation theorem implies that the $L^1(\mu) \to L^1_{weak}(\mu)$-boundedness proved in Theorem 3.3 is sufficient to derive $L^p(\mu) \to L^p(\mu)$ boundedness for $1 < p < \infty$, as follows.

Theorem 3.6. If T is an integral operator with integral kernel K satisfying (40, 41), and if T is bounded from $L^2(\mu)$ to $L^2(\mu)$, then T extends to a bounded operator from $L^p(\mu)$ to $L^p(\mu)$ for all $p \in]1, \infty[$.

Proof.

1. Theorem 3.3 proves that T is bounded from $L^1(\mu)$ to $L^1_{weak}(\mu)$; by Marcinkiewicz's theorem, T extends to a bounded operator from $L^p(\mu)$ to $L^p(\mu)$ for $1 < p \leq 2$.

2. For the range $2 \leq p < \infty$, we use the adjoint T* of T, defined by

$$\int (T^*\varphi)(\omega)\left(\overline{\psi(\omega)}\right) \mu(d\omega) = \int \varphi(\omega)\left(\overline{(T\psi)(\omega)}\right) \mu(d\omega).$$

It is associated with the integral kernel $K_0(\omega, \xi) = \left(\overline{K(\xi, \omega)}\right)$, which also satisfies the conditions (40, 41). This adjoint operator T* is exactly the adjoint operator used in operator theory on Hilbert spaces. It follows from Theorem 3.3 that T* is bounded from $L^1(\mu)$ to $L^1_{weak}(\mu)$, and hence by Theorem 3.5, that it is bounded from $L^p(\mu)$ to $L^p(\mu)$ for $1 < p \leq 2$. Since for $1/p + 1/q = 1$, T*: $L^p(\mu) \to L^p(\mu)$ is the adjoint operator of T: $L^q(\mu) \to L^q(\mu)$, it follows that T is bounded for $2 \leq q < \infty$. More explicitly, for readers unfamiliar with adjoints on Banach spaces,

$$\|T\varphi\|_q = \sup_{\psi \in L^p,, \|\psi\|=1} \left| \int (T\varphi)(\omega)\left(\overline{\psi(\omega)}\right) \mu(d\omega) \right| \text{ (if } 1/p + 1/q = 1)$$

$$= \sup_{\psi \in L^p, \|\psi\|=1} \left| \int \left(\int \varphi(\xi) K(\omega, \xi) \left(\overline{\psi(\omega)}\right) \mu(d\xi) \right) \mu(d\omega) \right|$$

$$= \sup_{\psi \in L^p, \|\psi\|=1} \left| \int \varphi(\xi)(T^*\psi)(\xi)\, \mu(d\psi) \right|$$

$$\leq \sup_{\psi \in L^p, \|\psi\|=1} \|\varphi\|_{L^q(\mu)} \ \|T^*\psi\|_{L^p(\mu)} \leq C\|\varphi\|_{L^q(\mu)}.$$

We now apply this to prove that a Wiener functional has some decay and some regularity and if the $\Psi_{j,k}(\omega) := 2^{-j/2}(2^{-j}\omega - ke_0), j, k \in \mathbb{Z}, \omega \in \mathbf{W}$, and where e_0 belongs to the Cameron–Martin space **H**, constitute an orthonormal basis for $L^2(\mu)$, then the $\Psi_{j,k}$ also provide unconditional bases for $L^p(\mu)$, $1 < p < \infty$.

What we need to prove is that if $\varphi = \sum_{j,k \in \mathbb{Z}} c_{j,k} \Psi_{j,k} \in L^p(\mu)$, then $\sum_{j,k \in \mathbb{Z}} \varepsilon_{j,k} c_{j,k} \Psi_{j,k} \in L^p(\mu)$ for any choice of the $\varepsilon_{j,k} = \pm 1$ (see Preliminaries in [8]).

We will assume that ψ is continuously Malliavin differentiable on the lattice space $(\mathbf{W}, |.|, \leq)$ where $|.|$ denotes the sup norm, and that both ψ, $\nabla\psi$ decay faster than $(1+|\omega|)^{-1}$:

$$|\psi(\omega)|, \|\nabla\psi(\omega)\|_{\mathbf{H}} \leq C(1+|\omega|)^{-1-\varepsilon}. \tag{45}$$

Then $\Psi \in L^p(\mu)$ for $1<p<\infty$, and $\varphi = \sum_{j,k} c_{j,k}\psi_{j,k}$ implies that $c_{j,k} = \int_{\mathbf{H}} \varphi(\omega)\psi_{j,k}(\omega)\,\mu(d\omega)$ because of the orthonormality of the $\psi_{j,k}$. We therefore want to show that, for aby choice of the $\varepsilon_{j,k} = \pm 1$, T_ε defined by

$$T_\varepsilon\varphi = \sum_{j,k} \varepsilon_{j,k}\left\langle \varphi,\ \psi_{j,k}\right\rangle_{L^2(\mu)}\psi_{j,k}$$

Is a bounded operator from $L^p(\mu)$ to $L^p(\mu)$. First, we know that T_ε is bounded from $L^2(\mu)$ to $L^2(\mu)$, since, denoting by $\|\ .\ \|_\mu$ (resp. $<.,.>_\mu$) the norm (resp. the scalar product) in $L^2(\mu)$,

$$\left(\|T_\varepsilon\varphi\|_\mu\right)^2 = \sum_{j,k}\left|\varepsilon_{j,k}\left\langle \varphi, \psi_{j,k}\right\rangle_\mu\right|^2 = \sum_{j,k}\left|\left\langle \varphi, \psi_{j,k}\right\rangle_\mu\right|^2 = \left(\|\varphi\|_\mu\right)^2,$$

so the $L^p(\mu)$-boundedness will follow by Theorem 3.5 if we can prove that T_ε is an integral operator with kernel satisfying (40, 41). This is the content of the following lemma.

Lemma 3.7. Choose $\varepsilon_{j,k} = \pm 1$, and define $K(\omega,\xi) := \sum_{j,k}\varepsilon_{j,k}\psi_{j,k}(\omega)\left(\overline{\psi_{j,k}(\xi)}\right)$. Then there exists C > 0 so that

$$|K(\omega,\xi)| \leq C|\omega-\xi|^{-1}$$

and

$$\|\nabla_\omega K(\omega,\xi)\|_{\mathbf{H}} + \|\nabla_\xi K(\omega,\xi)\|_{\mathbf{H}} \leq C|\omega-\xi|^{-2}.$$

Proof.

1.
$$\begin{aligned}|K(\omega,\xi)| &\leq \sum_{j,k}\left|\psi_{j,k}(\omega)\right|\cdot\left|\psi_{j,k}(\xi)\right| \\ &\leq C\sum_{j,k} 2^{-j}\left(1+|2^{-j}\omega - ke_0|\right)^{-1-\varepsilon}\left(1+|2^{-j}\xi - ke_0|\right)^{-1-\varepsilon} \text{by (3.13)}.\end{aligned}$$

Find $j(0) \in \mathbb{Z}$ so that $2^{j(0)} \leq |\omega-\xi| \leq 2^{j(0)+1}$. We split the sum over j into two parts: $j<j_0$ and $j \geq j(0)$.

2. $\sum_k (1+|a-k|)^{-1-\varepsilon}(1+|b-k|)^{-1-\varepsilon}$ is uniformly bounded for all values of $a, b \in \mathbb{R}$: in fact,

$$\sum_k (1+|a-k|)^{-1-\varepsilon}(1+|b-k|)^{-1-\varepsilon} \leq \sum_k (1+|a-k|)^{-1-\varepsilon}$$
$$\leq \sup_{0\leq a'\leq 1} \sum_k (1+|a'-k|)^{-1-\varepsilon} \leq 2 \sum_{l\in\mathbb{N}} (1+l)^{-1-\varepsilon} < \infty.$$

Hence,

$$\sum_{j\geq j(0)}\sum_k 2^{-j}(1+|2^{-j}\omega - ke(0)|)^{-1-\varepsilon}(1+|2^{-j}\xi - ke(0)|)^{-1-\varepsilon}$$
$$\leq C\sum_{j\geq j(0)} 2^{-j} \leq C.2^{-j(0)+1} \leq 4C|\omega-\xi|^{-1}.$$

3. The part $j < j(0)$ is a little less easy.

$$\sum_{-\infty<j\leq j(0)-1} 2^{-j}\sum_k [(1+|2^{-j}\omega - ke(0)|).(1+|2^{-j}\xi - ke(0)|)]^{-1-\varepsilon}$$
$$= \sum_{j\geq -j(0)+1} 2^{j}\sum_k [(1+|2^{j}\omega - ke(0)|).(1+|2^{j}\xi - ke(0)|)]^{-1-\varepsilon}$$
$$\leq 2^{1+\varepsilon}\sum_{j\geq j(0)+1} 2^{j}\sum_k [(2+|2^{j}\omega - ke(0)|).(2+|2^{j}\xi - ke(0)|)]^{-1-\varepsilon}. \quad (46)$$

Find $k_0 \in \mathbb{Z}$ so that $k_0 \leq 2^j(|\omega| + |\xi|)/2 \leq k_0 + 1$ and define $l = k - k_0$. Then

$$2+|2^j\omega - ke(0)| = 2 + |2^j(\omega-\xi)/2 - le(0) + (2^j(\omega+\xi)/2 - k_0e(0))|$$
$$\geq 1 + |2^j(\omega-\xi)/2 - le(0)|;$$

similarly,

$$2+|2^j\xi - ke(0)| \geq 1 + |2^j(\xi-\omega)/2 - le(0)|.$$

Consequently, with $\omega_1 := 2^j(\omega-\xi)/2$,

$$\sum_k [(2+|2^j\omega - ke(0)|).(2+|2^j\xi - ke(0)|)]^{-1-\varepsilon}$$
$$\leq \sum_l [(1+|\omega_1 + le(0)|).(1+|\omega_1 + le(0)|)]^{-1-\varepsilon} \leq C(1+|\omega_1|)^{-1-\varepsilon},$$

so that

$$\leq C\sum_{j\geq -j(0)+1} 2^j(1+2^j|\omega-\xi|/2)^{-1-\varepsilon}$$
$$\leq C\sum_{j'\geq 1} 2^{j'-j(0)}\left(1+2^{j'-j(0)}(1/2)2^{j(0)+1}\right)^{-1-\varepsilon} \text{ as we have } |\omega-\xi| \leq 2^{j(0)+1}, \quad (47)$$
$$\leq C2^{-j(0)}\sum_{j'\geq 1} 2^{j'}\left(1+2^{j'}\right)^{-1-\varepsilon}$$
$$\leq C'2^{-j(0)} \leq 2C'|\omega-\xi|^{-1}.$$

It therefore follows that $|K(\omega,\xi)| \leq C|\omega-\xi|^{-1}$

4. For the estimates on $\nabla_\omega K$ and $\nabla_\xi K$, we write

$$\|\nabla_\omega K(\omega,\xi)\|_{\mathbf{H}} \leq \sum_{j,k} 2^{-j} \|\nabla\psi(2^{-j}\omega - ke_0)\|_{\mathbf{H}} |\psi(2^{-j}\omega - ke_0)|$$
$$\leq C\sum_{j,k} 2^{-2j}\left[(1+|2^{-j}\omega - ke_0|)(1+|2^{-j}\xi - ke_0|)\right]^{-1-\varepsilon}$$

and we follow the same technique; we obtain.
$\|\nabla_\omega K(\omega,\xi)\|_{\mathbf{H}}, \|\nabla_\xi K(\omega,\xi)\|_{\mathbf{H}} \leq C|\omega-\xi|^{-2}$.■

It therefore follows from the lines we write before the lemma the following theorem [20, 21].

Theorem3.8. If ψ is a Wiener functional continuously Malliavin differentiable and $|\psi(\omega)|, \|\nabla\psi(\omega)\|_{\mathbf{H}} \leq C(1+|\omega|)^{-1-\varepsilon}$, and if the $\psi_{j,k}(\omega) := 2^{-j/2}\psi(2^{-j}\omega - ke_0)$ constitute an orthonormal basis for $L^2(\mu)$, $e_0 \in \mathbf{H}$ being given, then the $\{\psi_{j,k}; j,k \in \mathbb{Z}\}$ also constitute an unconditional basis for all $L^p(\mu)$ – spaces, $1<p<\infty$.

1.4 Periodized Wiener wavelets

Even if L^1-spaces do not have unconditional bases, Wiener wavelets still outperform Fourier Wiener analysis in some sense. To illustrate this, let us first introduce " periodized Wiener wavelets ". Given a multiresolution Wiener analysis with scaling Wiener functional ϕ and Wiener wavelet ψ, both with reasonable decay (say, $|\phi(\omega)|,\ |\psi(\omega)| \leq C(1+|\omega|)^{-1-\varepsilon}$), we define

$$\phi_{j,k}^{per}(\omega) := \sum_{l\in\mathbb{Z}} \phi_{j,k}(\omega + le_0),\ \psi_{j,k}^{per}(\omega) := \sum_{l\in\mathbb{Z}} \psi_{j,k}(\omega + le_0);$$

and

$$\mathbf{V}_j^{per} := closure\ of\ span\left\{\phi_{j,k}^{per}; k\in\mathbb{Z}\right\},$$

$$\mathbf{W}_j^{per} := closure\ of\ span\left\{\psi_{j,k}^{per}; k\in\mathbb{Z}\right\}$$

First, notice that we have: $\sum_{l\in\mathbb{Z}}\phi(\omega+le_0) = 1$. In fact, put $\varphi(\omega) := \sum_{l\in\mathbb{Z}}\phi(\omega+le_0)$. The conditions on ϕ (continuity and its " reasonable decay ") ensure that φ is well defined and continuous. Moreover, we can write:

$$\phi(\omega) = \sum_{n\in\mathbb{Z}} c_n \phi(2\omega - ne_0) \text{ with } (c_n)_n \in \ell^2(\mathbb{Z}).$$

Then,

$$\varphi(\omega) = \sum_l\sum_n c_n\phi(2\omega - 2le_0 - ne_0) = \sum_l\sum_m c_{m-2l}\phi(2\omega - me_0)$$
$$= \sum_m\left(\sum_j c_{m-2j}\right)\phi(2\omega - me_0) = \sum_m \phi(2\omega - me_0) = \varphi(2\omega).$$

Hence, φ is continuous, periodic with period e_0, and

$$\varphi(\omega) = \varphi(2\omega) = \ldots = \varphi(2^n\omega) = \ldots$$

It follows that φ is a constant Wiener functional. We then put:

$$\sum_l \phi(\omega - le_0) = c$$

Since
$\int \phi \, d\mu = 1$, this constant is necessarily equal to 1. We deduce, for $j \geq 0$, $\phi_{j,k}^{per}(\omega) = 2^{-j/2} \sum_l \phi\left(2^{-j}\omega - ke_0 + 2^{-j}le_0\right) = 2^{j/2}$, so that the $\mathbf{V}_j^{per}$, for $j \geq 0$, are all identical one-dimensional spaces, containing the constant functionals.

Similarly, $\sum_l \psi(\omega + (l/2)e_0) = 0$. In fact, $\sum_l \psi(\omega + (l/2)e_0) = \sum_l \sum_n (-1)^n c_{-n+1} \phi(2\omega + le_0 - ne_0)$ where the c_n are the coefficients appearing in (13), that is,

$$c_n = \langle \phi, \phi_{-1,n} \rangle = \sum_{k,m} (-1)^{m+1} c_m \phi(2\omega + ke_0)(k = l - n, m = -n + 1)$$
$$= 0 \left(\text{because } \sum_m c_{2m} = \sum_m c_{2m+1}\right).$$

Hence, for $j \geq 1 \geq$, $\mathbf{W}_j^{per} = \{0\}$. We therefore restrict our attention to the $\mathbf{V}_j^{per}$, $\mathbf{W}_j^{per}$ with $j \leq 0$. Obviously $\mathbf{V}_j^{per}, \mathbf{W}_j^{per} \subset \mathbf{V}_{j-1}^{per}$, a property inherited from the non-periodized spaces. Moreover, $\mathbf{W}_j^{per}$ is still orthogonal to $\mathbf{V}_j^{per}$, because

$$\int_{\mathbf{H}} \Psi_{j,k}^{per}(\omega) \phi_{j,k'}^{per}(\omega)\, \mu(d\omega) = \sum_{l,l' \in \mathbb{Z}} 2^{-j} \int_{\mathbf{H}} \psi\left(2^{-j}\omega + 2^{-j}le_0 - ke_0\right) \left(\overline{\phi\left(2^{-j}\omega + 2^{-j}l'e_0 - k'e_0\right)}\right) \mu(d\omega)$$
$$= \sum_{l,l' \in \mathbb{Z}} 2^{|j|} \int_{]l'e_0, (l'+1)e(0)]} \Psi\left(2^{|j|}\omega + 2^{|j|}(l - l')e_0 - ke_0\right) \overline{\phi\left(2^{|j|}\xi - k'e_0\right)}\, \mu(d\xi),$$

because $j \leq 0$,

$$= \sum_{r \in \mathbb{Z}} \left\langle \psi_{j,k+2^{|j|}r}, \phi_{j,k'} \right\rangle_\mu = 0.$$

It follows that, as in the non-periodized case, $\mathbf{V}_{j-1}^{per} = \mathbf{V}_j^{per} \oplus \mathbf{W}_j^{per}$. The spaces $\mathbf{V}_j^{per}$, $\mathbf{W}_j^{per}$ are all finite-dimensional: since $\Phi_{j,k+m2^{|j|}} = \Phi_{j,k}$ for $m \in \mathbb{Z}$, and the same is true for Ψ, both $\mathbf{V}_j^{per}$ and $\mathbf{W}_j^{per}$ are spanned by the $2^{|j|}$ Wiener functionals obtained from $k = 0, 1, \ldots, 2^{|j|} - 1$. These $2^{|j|}$ Wiener functionals are moreover orthonormal; in, for example, $\mathbf{W}_j^{per}$ we have, for $0 \leq k, k' \leq 2^{|j|} - 1$,

$$\left\langle \Psi_{j,k}^{per}, \Psi_{j,k'}^{per} \right\rangle_\mu = \sum_{r \in \mathbb{Z}} \left\langle \psi_{j,k+2^{|j|}r}, \; \psi_{j,k'} \right\rangle_\mu = \delta_{k;k'}.$$

We have therefore a ladder of multiresolution spaces, $\mathbf{V}_0^{per} \subset \mathbf{V}_{-1}^{per} \subset \mathbf{V}_{-2}^{per} \subset \ldots$, with successive orthogonal complements $\mathbf{W}_0^{per}$ (of $\mathbf{V}_0^{per}$ in $\mathbf{V}_{-1}^{per}$), $\mathbf{W}_1^{per}$, ..., and orthonormal bases $\left\{\phi_{j,k}; k = 0, \ldots, 2^{|j|} - 1\right\}$ in $\mathbf{V}_j^{per}$, $\left\{\psi_{j,k}; k = 0, \ldots, 2^{|j|} - 1\right\}$ in $\mathbf{W}_j^{per}$. Since the closed space spanned by $\overline{\bigcup_{j \in -\mathbb{N}} \mathbf{V}_j^{per}}$ is $L^2(\mu)$ (this follows from the corresponding non-periodized version), the Wiener functionals in $\left\{\phi_{0,0}^{per}\right\} \cup \left\{\psi_{j,k}^{per}; -j \in \mathbb{N}, k = 0, \ldots, 2^{|j|} - 1\right\}$ constitute an orthonormal basis in $L^2(\mu)$. We will relabel this basis as follows:

$$\begin{aligned}
&\psi_0(\omega) = 1 = \phi_{0,0}^{per}(\omega), \psi_1(\omega) = \psi_{0,0}^{per}(\omega), \psi_2(\omega) = \psi_{-1,0}^{per}(\omega), \\
&\psi_3(\omega) = \psi_{-1,1}^{per}(\omega), \ldots, \\
&\psi_{2^j}(\omega) = \psi_{-j,0}^{per}(\omega), \ldots, \psi_{2^j+k}(\omega) = \psi_{-j,k}^{per}(\omega) = \psi_{2^j}\left(\omega - k2^{-j}e_0\right) \\
&\text{for } 0 \le k \le 2^j - 1, \ldots .
\end{aligned}$$

Then this basis has the following property.

Theorem 4.1. If φ is a continuous periodic Wiener functional with period e_0, then there exist $\alpha_n \in \mathbb{C}$ so that

$$\lim_N \left\| \varphi - \sum_{n=0, \ldots, N} \alpha_n \psi_n \right\|_\infty = 0, \tag{48}$$

where $\|.\|_\infty$ denotes the norm of $L^\infty(\mu)$.

Proof.

1. Since the ψ_n are orthonormal, we necessarily have $\alpha_n = \langle \varphi, \psi_n \rangle_\mu$. Define

$$S_N \varphi = \sum_{n=0, \ldots, N} \langle \varphi, \psi_n \rangle_\mu .$$

In a first step we prove that the S_N are uniformly bounded, that is,

$$\|S_N \varphi\|_\infty \le C \|\varphi\|_\infty, \tag{49}$$

with C independent of φ or N.

2. If $N = 2^j$, then S_{2^j} is the orthogonal projection operator on $\mathbf{V}_{-j}^{per}$; hence

$$(S_{2^j}\varphi)(\omega) = \sum_{k=0, \ldots, 2^{|j|}-1} \left\langle \varphi, \phi_{-j,k}^{per} \right\rangle_\mu \phi_{-j,k}^{per}(\omega) = \int_{\mathbf{H}} K_j(\omega, \xi) \varphi(\xi)\, \mu(d\xi),$$

with $K_j(\omega, \xi) = \sum_{k=0, \ldots, 2^{|j|}-1} \phi_{-j,k}^{per}(\omega) \left(\overline{\phi_{-j,l}^{per}(\xi)}\right)..$

Consequently,

$$\|S_2 \varphi\|_\infty \le \left(\sup_{\omega \in \mathbf{H}} \int_{\mathbf{H}} |K_j(\omega, \xi)|\; \mu(d\xi) \right) \|\varphi\|_\infty .$$

Now,

$$\sup_{\omega\in\mathbf{H}}\int_{\mathbf{H}}|K_j(\omega,\xi)|\,\mu(d\xi)$$

$$\leq \sup_{\omega\in\mathbf{H}}\int_{\mathbf{H}}\sum_{k=,,2^{|j|}-1}\sum_{l,l'\in\mathbb{Z}}\left|\phi_{-j,k}(\omega+le_0)\right|.\left|\phi_{-j,k}(\xi+l'e_0)\right|\,\mu(d\xi)$$

$$\leq \sup_{\omega\in\mathbf{H}}\int_{\mathbf{H}}\sum_{k=0,\,\dots,\,2^{|j|}-1}\sum_{l\in\mathbb{Z}}2^j\left|\phi(2^j(\omega+le_0)-ke_0)\right|.\left|\phi(2^j\xi-ke_0)\right|\,\mu(d\xi)$$

$$\leq C\sup_{\omega'\in\mathbf{H}}\sum_{k=0,\,\dots,\,2^{|j|}-1}\sum_{l\in\mathbb{Z}}|\phi(\omega'+2^jle \leq C\sup_{\omega'\in\mathbf{H}}\sum_{m\in\mathbb{Z}}|\phi(\omega'+me_0)|,$$

and this is uniformly bounded if $|\phi(\omega)|\leq C(1+|\omega|)^{-1-\varepsilon}$. This establishes (49) for $N=2^j$.

3. If $N=2^j+m$, $0\leq m\leq 2^j-1$, then

$$(S_N\varphi)(\omega)=\left(S_2^j\varphi\right)(\omega)+\sum_{k=0,\,\dots,\,m}\left\langle\varphi,\psi^{per}_{-j,k}\right\rangle_\mu\psi^{per}_{-j,k}(\omega).$$

Estimates exactly similar to those in point 2 show that the $L^\infty(\mu)$-norm of the second sum is also bounded by $C\|\varphi\|_\infty$, uniformly in j, which proves (49) for all N.

4. Take now $\varphi\in\mathbf{E}=\bigcup_{j\in-\mathbb{N}}\mathbf{V}^{per}_j$. Then $\varphi\in\mathbf{V}^{per}_{-J}$ for some $J>0$,

so that $\left\langle\varphi,\psi^{per}_{-j,k}\right\rangle_\mu=0$ for $j'\geq J$, i.e., $\langle\varphi,\psi_l\rangle_\mu=0$ for $l\geq 2^J$. Consequently, $\varphi=S_N\varphi$ if $N\geq 2^J$, so that (48) clearly holds. Since **E** is dense in the space of continuous periodic Wiener functionals equipped with the $\|\ .\ \|_\infty$-norm, the theorem follows.∎.

We deduce a similar theorem for $L^1(\mu)$.

Theorem 4.2. If $\varphi\in L^1(\mu)$, then $\lim_N\left\|\varphi-\sum_{n=0,\,\dots,\,N}\langle\varphi,\psi_n\rangle_\mu\psi_n\right\|_\mu=0$.

Proof.
As we have the following:

$$\|\varphi\|_{L^1(\mu)}=\sup\left\{\left|\langle\varphi,\psi\rangle_\mu\right|/\psi\text{ continuous, periodic with period }e_0,\|\psi\|_\infty\leq 1\right\},$$

this leads immediately to

$$\begin{aligned}\|S_N\varphi\|_{L^1(\mu)}&=\sup\left\{\left|\langle S_N\varphi,\psi\rangle_\mu\right|\psi\text{ continuous, }e_0\text{–periodic, }\|\psi\|_\infty\leq 1\right\}\\&=\sup\left\{\left|\langle\varphi,S_N\psi\rangle_\mu\right|\psi\text{ continuous, }e_0\text{–periodic, }\|\psi\|_\infty\leq 1\right\}\\&\leq C\|\varphi\|_{L^1(\mu)}\end{aligned}\tag{50}$$

by the uniform bound (49) and because $\left|\langle\varphi,\psi\rangle_\mu\right|\leq\|\varphi\|_{L^1(\mu)}\|\psi\|_\infty$.

Since $\mathbf{E} = \cup_{j \in -\mathbb{N}} \mathbf{V}_j^{per}$ is dense in $L^1(\mu)$, the uniform bound (50) is sufficient to prove the theorem.∎.

Remark. The ordering of the ψ_n is important in Theorems 4.1 and 4.2: we have a Schauder basis, but not an unconditional basis.

Author details

Claude Martias
French West Indies University, French West Indies

*Address all correspondence to: claude.martias0157@orange.fr

References

[1] Martias C. The wavelet transform for Wiener functionals and some applications. Stochastics: An International Journal of Probability and Stochastic Processes. 2014. DOI: 10.1080/17442508.2013.879144

[2] Argawal A, Maheswaran R, Marwan N, Caesar L, Kurths J. Wavelet – Based multiscal similarity measure for complex networks. The European Physical Journal B. 2018;**91**:296. DOI: 10.1140/epjb/e2018-90460-6

[3] Ghaderpour E, Pagiatakis SD, Hassan QK. Une enquête sur la détection des changements et l'analyse des séries chronologiques avec des applications. Sciences appliquées. 2021;**11**(13):6141. DOI: 10.3390/app11136141

[4] Hongia Q, Tiantian L. Chen Genda (2019 – 01 – 01), Transformations d'ondelettes adaptatives pressées par synchro avec des paramètres optimaux pour les séries chronologiques arbitraires. Systèmes mécaniques et traitement du signal. 2018. Bibcode: 2019MSSP. 366Q;**114**:366-377. DOI: 10.1016/j.ymssp. 2018.05.020

[5] Mallat S. Multiresolution approximations and wavelet orthonormal bases of $L^2(\mathbb{R})$. Transactions of the American Mathematical Society. 1989;**315**: 69-87

[6] Goodman TNT, Lee SL, Tang WS. Wavelet in wandering subspaces, Transactions of the American Mathematical Society. 1993;338(N°2)

[7] Halmos PR. A Hilbert Space Problem Book. 2nd ed. Springer – Verlag; 1982

[8] Daubechies I. Ten Lectures on Wavelets, Society for Industrial and Applied Mathematics. Philadelphia, PA: SIAM; 1992

[9] Haar A. Zur Theorie der orthogonalen Funktionen – Système. Mathematische Annalen. 1910;**69**:331-371

[10] Meyer PA, Yan JA. In: Séminaire de Proba XXV, Azema J, Meyer PA, Yor M, editors. Les « fonctions caractéristiques » des distributions sur l'espace de Wiener. Berlin, Heidelberg: Springer – Verlag; 1991

[11] Lee SL. Fourier transforms of B – Splines and fundamental splines for cardinal Hermite interpolations. Proceedings of the American Mathematical Society. 1976;**57**:291-296

[12] Meyer Y. Ondelettes, fonctions splines et analyses graduées. Italy: Lectures given et the University of Torino; 1986

[13] Meyer Y. Ondelettes et Opérateurs, I: Ondelettes, II: Opérateurs de Calderon – Zygmund, III: Opérateurs Multilinéaires. Paris: Hermann; 1990

[14] Shoenberg IJ. Cardinal Spline Interpolation, CBMS – NSF Series in Appl. Philadelphia, Pa: Math. n°12, SIAM; 1973

[15] Nualart D. The Malliavin Calculus and Related Topics. 2nd ed. Paris: Springer – Verlag; 2006

[16] Robertson JB. On wandering subspaces for unitary operators. Proceedings of the American Mathematical Society. 1965;**16**:233-236

[17] Daubechies I. The wavelet transform, time – Frequency localization and signal analysis. IEEE Transactions on Information Theory. 1990;**36**:961-1005

[18] Duffin RJ, Schaeffer AC. A class of nonharmonic Fourier series. Transactions of the American Mathematical Society. 1952;72:341-366

[19] Stein E, Weiss G. Introduction to Fourier Analysis on Euclidean Spaces. Princeton University Press; 1971

[20] Üstünel AS. An Introduction to Analysis on Wiener Space, Lect. Notes in Math. 1610. Springer; 1995

[21] Young RM. An Introduction to Nonharmonic Fourier Series. New York: Academic Press; 1980

Section 3

Wavelet Transforms in Spectral Sensing, Monitoring Condition and Restoration of Signal

Chapter 5

Wavelet Transform-Spectrum Sensing

Himanshu Monga, Dikshant Gautam and Saksham Katwal

Abstract

Spectrum sensing is a vital cognitive radio function that protects licensed users from dangerous interference and finds accessible spectrum for better spectrum use. In practice, however, multipath fading, shadowing, and receiver uncertainty frequently degrade detection performance. Communication performance and continuity in cognitive radio networks are heavily dependent on how well the spectrum sensing function is implemented. The significance of selecting the right wavelet system is discussed.

Keywords: cognitive radio, spectrum sensing, multipath fading, shadowing, wavelet system

1. Introduction

Spectrum sensing is crucial in cognitive radio technology for effective bandwidth usage. The problem of interference between adjacent spectrums can be efficiently avoided by carefully finding the spectrum boundaries. Increase the data rates in the channel by adding bands and increasing the data rates in the channel. Various attempts have been made. Wavelet Edge detection is a technique for accurately detecting spectral boundaries [1]. The method that is extensively used nonetheless, the system's efficiency is quite low. Depending on the wavelet type utilized. After analyzing the nature of peaks in the power spectral density of the spectrum, a spectrum sensing technique selects the best wavelet function for the supplied spectrum. For the problem of edge detection, traditional solutions use a specific wavelet function, followed by energy detection or periodicity detection [2, 3]. However, a single wavelet function cannot be employed efficiently for real-time spectrum data with significantly different PSD properties.

1.1 Spectrum sensing

Cognitive Radio (CR) is a new technology that aims to solve the problem of wasteful spectrum use by increasing idle spectrum usage in both time and space. CR has the capacity to dynamically access the spectrum, determining which frequencies are not in use, and reserving them for data transmission and reception. When compared to traditional radio techniques, cognitive radio has a number of advantages, such as the ability to activate several licensed frequency bands to allow an unconstrained secondary user to communicate with another CR in some spectrum policy that defines some CR rules and limitations, and the ability to transmit data simply by changing the operating factor without any changes to the hardware components [4, 5]. Spectrum sensing is a vital step in the evolution of technology Cognitive Radio, in which the major users are detected in that spectrum band to detect spectrum holes and minimize unintended interference. Spectrum sensing can be done in a variety of ways. The following three types of spectrum sensing techniques are based on primary user availability.

Non-cooperative detection (Transmitter detection), cooperative detection, and interference detection are the three types of detection. Wavelet-based spectrum sensing algorithms are utilized for signal edge detection in transmitter detection, where primary users are present to detect spectrum opportunities.

1.2 Wavelet transform

The wavelet theory is used to evaluate signals by breaking them down into their constituents and basic functions. The wavelet transform can characterize the local regularity of signals and is a mathematical tool for analyzing singularities and irregular structures. In order to study the primary users, the wavelet transform approach for spectrum detection in CR is well justified. Wavelets are useful for studying fluctuations in signals and spectrum since they are described by both scale and position. The concept of local regularity is used to convey scale, while a list of domains is used to describe time aspects. The Continuous Wavelet Transform (CWT) is a two-parameter wavelet function signal expansion [6].

When compared to other types of spectrum sensing techniques, this wavelet transform technique takes much less time to detect whether the principal user is consuming the spectrum or not. When the de-noising and compression processes are at their center points, the disintegration is considered complete. Because only the available frequency, i.e. spectrum holes in which the secondary user can communicate, is denoted at each level of the wavelet decomposition.

2. Traditional spectrum sensing methods

Spectrum sensing is essentially a form of energy detection in which the presence or absence of useful data in a specific frequency band is determined. The presence of data indicates a rise in energy from the noise floor, or the presence of a noise-like signal with higher-order periodicity. Traditional techniques work in narrowband and use a set of FIR filters that are tuned for the frequency range in question. While this strategy works well for narrow bandwidths, it becomes inefficient as the dynamic range of Cognitive Radios' functioning expands. This approach is quite difficult for ultra-wideband cognitive radio systems. However, it is recommended that the reader has first-hand knowledge of traditional methods in order to comprehend the properties of Wavelet-based algorithms so that the discussion is limited to those parts of spectrum sensing that cannot be done using these methods.

2.1 Short-time fourier transform (STFT)

The "Windowed Fourier Transform" is another name for this approach. In general, the Fourier Transform is a time-to-frequency domain transformation that yields time averaged values for distinct frequency components. While this technique is beneficial for examining frequency components, it does not allow us to determine when they occur. i.e. frequency localization in time is impossible. The entire signal is separated into smaller segments (using an appropriate window function) and the Fourier Transform is obtained for these intervals in a short-time Fourier transform.

While this method outperforms Fourier analysis in terms of results, it suffers from limited frequency resolution, large volatility in the predicted power spectrum, and high side lobes/leakages.

2.2 Periodogram

Until better methods were found to replace it, the Periodogram was one of the most widely utilized Spectrum Sensing techniques. The infinite length sequence is trimmed with a rectangular window function in this manner, and the FFT is obtained. The square of the FFT yields a rough Spectral Density plot. The signals are abruptly truncated, which is a fundamental flaw in this strategy. This produces a Dirichlet Kernel in the frequency domain, which is defined by the width of the main lobe and side lobes, as detailed in [7, 8]. As a result, spectral leakage occurs at the discontinuities. Furthermore, time-frequency localization was not possible with this strategy.

2.3 Matched filter approach

This is a test method of detection. This is the quickest way to detect spectrum, but it fails because it requires prior knowledge of the primary user's modulation type, pulse shaping, and packet format. Coherence requires precise timing and synchronization. However, temporal dispersion and Doppler shifts can occur as a result of channel fading effects, affecting synchronization. In [9] there is a lot of information about how to implement matching filters.

2.4 Cyclo-stationary feature detection

First-order periodicity can be applied to any observable regularity. Because of modulation techniques or in the sent data. The signal gains a specific periodicity as a result of source coding, which can be useful. Only nonlinear time-invariant transformations of the time can be seen. Series [10]. Cyclostationarity is the name for this form of second-order periodicity. The mean, autocorrelation, and other statistical properties, in general, demonstrate a pattern of behavior This can be used to determine whether something is present or not a frequency band's worth of data. The advantage is that this is the only method that accurately measures spectral occupancy in bands with very low SNR. The borders, on the other hand, cannot be precisely specified. As a result, this technique, in combination with border detection, is frequently utilized in the development of algorithms. Adoum and Jeoti [2] provides a full explanation of how to compute the Cyclic Spectrum Density.

2.5 Multi taper spectrum estimation

In this strategy, the issues that plagued the Periodogram approach are partly mitigated. Multiple orthogonal filters are used to limit spectral leakage and volatility in the calculated power. Consider the signal $X(n) = [x(n)\ x(n-1)\ x(n-2)\ \ldots\ x(n-M+1)]\ T$, which consists of M samples. These data points are used to create an orthogonal basis, and the expansion coefficients are modified using a set of values that represent the way the spectrum tapers. Thomson [11] provides a more thorough formulation. The average of numerous Periodograms with different windows can be deduced as MTSE. As a result, each window shape displays distinct aspects of the spectrum, while the average value smooths out the discontinuous points and reduces spectral leakage. This solution suffers from expensive computations and the fact that it cannot totally solve the Spectrum leakage problem, although better methods can.

2.6 Quadrature mirror filter banks

The entire wideband spectrum is separated into M-bands using this strategy. These are predefined bands in which user activity is monitored using n stages of Quadrature Mirror Filters tuned for the band, with $M = 2n$. As a result, as with the Matched Filter Method, previous knowledge of the Primary user is required. However, because a set of Filter banks is utilized, the same filters can be used for data reception after spectral gaps have been detected [12]. As a result, they serve a dual purpose. In addition, the tree structure aids in the reduction of computer complexity. The energy detection process starts with only two filters in the first step. Thresholds are set to determine whether or not to move on to the next step. The procedure of analyzing the sub-bands is skipped if the signal energy is larger than the threshold. The difficulty with this strategy is that free spectra will be wasted if the spectrum's border does not correspond with the designated pass-band of a particular filter. As a result, spectral holes narrower than the pass-band of the last stage filter are not detectable using this technique. Furthermore, the channel fading effects have a significant impact on its performance.

These methodologies, as well as their faults, prompted researchers to develop a domain transformation technique that could analyze both spatial and spectral data at the same time. Wavelets developed as a promising concept with a lot of promise for fixing the challenges described above. To a large extent, they helped with temporal frequency localization. Their ability to respond to function discontinuities (singularity) also enables researchers to use them in a variety of border detecting applications. The mathematics of Wavelet Edge Detection is presented in the next section.

3. Wavelet theory of edge detection

Before getting into the analysis of Edge Detection with Wavelets, it's a good idea to give a quick overview of how the Wavelet theory came to be. This is followed by the Edge Detection Technique in the following subsection.

Edges in the frequency spectrum are generated when a signal's frequency changes. This feature was utilized in the detection process. The wavelet transform is commonly used on these sub-bands for detection and estimation of native spectral irregularities/transitions, which carries significant information on Power Spectral Densities (PSDs) and frequency locations [13]. The entire frequency range is divided into various sub-bands, and the wavelet transform is generally used on these sub-bands for detection and estimation of native spectral irregularities/transitions, which carries significant information on PSDs and frequency locations. The wavelet transform has been employed instead of the standard Fourier transform because it provides information about the exact placement of different frequency locations and spectral densities. Furthermore, the Fourier transform can only display the various frequency components, not their location. Edges representing transitions from empty to the occupied band have been sought for in the spectral densities of all sub-bands. The initial stage in wideband spectrum sensing is to determine the frequency position of each channel in the RF spectrum. Regardless of the PSD's actual shape, sharp variation points (singularities) along the channel's edges are expected [14]. The difficulty of detecting these anomalies might be thought of as an edge detection problem. Depending on the PSD level of each channel, the discovered spectrum bands have now been categorized as occupied or unoccupied. The existence of an edge indicates that the band contains PU.

PSD of the received signal can be written as:

$$(f) = \sum(f) + (f), f\ [f0, fN] \tag{1}$$

Where (f) & (f) are the signal and noise PSDs, respectively, within the nth band, and (f) & (f) are the signal and noise PSDs, respectively. The wideband frequency range is denoted by $f\ [f0, fN]$. This technique has worked well for ultra-wideband (3–10GHz) CRs with a variety of narrowband incumbents and other users like as WiMAX and Wi-Fi. Pros: Adapts quickly to changing PSD structures. Cons: Characterizing the whole bandwidth was required at higher sampling rates.

4. Types and classification of wavelets

Wavelets are divided into two groups, as seen in **Figure 1** [15]. Wavelets, which are characterized by mathematical formulas that are continuous and infinite in nature, are the first category [15]. They are also known as crude wavelets because they must be converted to wavelet filters with a finite number of discrete points before they can be used in any signal processing system. The Mexican hat wavelet (**Figure 1**) is a good illustration of this sort of wavelet.

Wavelets that start off as filters with two points of definition in the initial state are the second category of wavelets [15]. These wavelets form an approximation of a

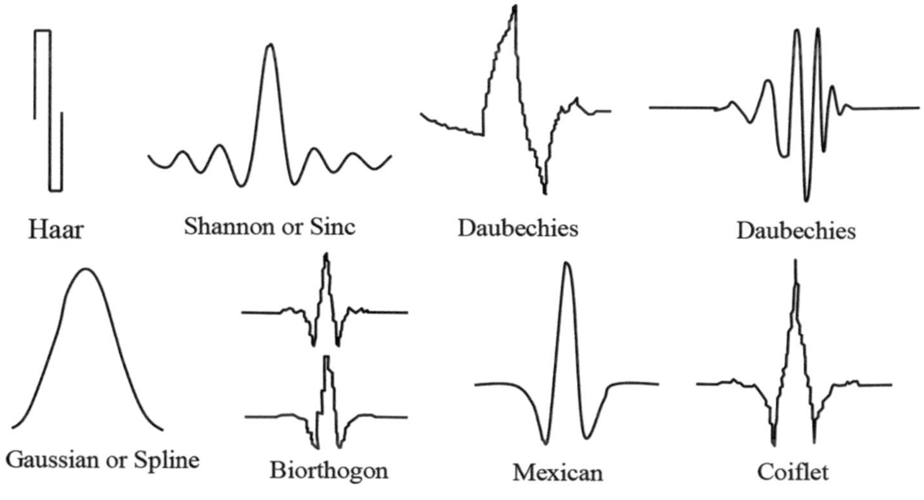

Figure 1.
Types of Wavelet.

continuous wavelet by interpolating and extrapolating more points from the starting two points. The Daubechies 4 (Db4) wavelet, depicted in **Figure 1** [15, 16], is a good example of this type of wavelet.

Wavelets are widely used in a variety of scientific and technical sectors. They each have different features that make them best suited for a specific use. The wavelet and scaling functions of biorthogonal wavelets are symmetric. Because the human vision system is more tolerant of symmetric faults than asymmetric errors, this characteristic makes biorthogonal wavelets suitable for human vision perception [17]. Wavelet transforms such as Haar, symlet, and Daubechies are known to have asymmetric errors.

Shannon wavelets feature a sharp localization and limitless frequency support. As a result, they're great for identifying events with specified frequencies [14, 18]. Two key characteristics of Haar wavelets make them perfect for edge identification. The first is that Haar wavelets maintain signal energy [19]. The second virtue is that Haar wavelets are exactly reversible without edge effects [20], which is critical for edge detection.

Coiflet wavelets are ideal in numerical analysis because of their high number of vanishing moments and almost interpolating and linear phase low-pass within a given passband [21]. They can also deal with fractals in signal processing because of their almost interpolating and linear phase low-pass within a given passband.

The Daubechies wavelets are characterized by a high degree of regularity, many vanishing moments, and approximate symmetry. Because the wavelets get smoother as the vanishing moments grow, these characteristics are extremely desirable in signal processing and data compression applications [22, 23]. In Section 6, an example of Daubechies applicability for spectrum sensing is given.

The Morlet wavelet is frequently used in signals that are related to the environment, such as seismic vibrations. The Morlet wavelet's ability to capture both amplitude and phase features of a signal while maintaining the signal's temporal aspect makes it appealing in this application [24–26].

Patch and gap events are well-localized in Mexican hat wavelets. They also have a lot in common with MUAPs (motor unit action potentials), hence they can be used in EMG (electromyography) [27, 28].

5. Wavelet-based spectrum sensing

Wavelets are short-duration signals. Wavelets differ from sinusoids in theory because, whereas sinusoids stretch from—to, wavelets have a finite beginning and ending locations. The premise in wavelet-based spectrum sensing is that the CR system receives a signal that spans N spectrum bands and that the CR must detect the PSD (power spectrum density) levels and frequency positions of each band. **Figure 2** [29] depicts a spectrum band between $f0$ and fN, with sub-band frequencies at $f0, f1$, and fn. Shiann-Shiun et al. [30] defines the nth band in **Figure 2** as follows:

$$Bn : \{ f \in Bn : fn - 1 \leq f < fn\}, n = 1, 2, \ldots, N. \quad (2)$$

PSD of a CR system receiving an input signal:

$$Sr(f) = \sum_{n=1}^{N} \alpha 2nSn(f) + S\omega(f), f \in [f0, fN] \quad (3)$$

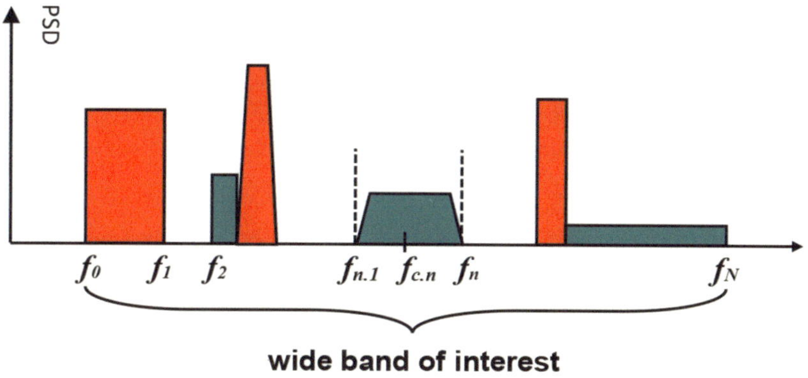

Figure 2.
N frequency bands with piecewise smooth PSD.

There are three different ways to apply the wavelet approach to spectrum sensing. The continuous wavelet technique, discrete wavelet technique, and discrete wavelet packet technique are examples of these. DWT (Discrete Wavelet Transform) and CWT (Continuous Wavelet Transform) techniques are also used for reduced BER with an increase in SNR in optical wireless systems which are based on On–Off keying modulation (OOK) [31].

5.1 Continuous wavelet transform-based spectrum sensing

The spectrum sensing based on continuous wavelet (CWT) [29] assesses the similarity between a signal and an analysis function. by making use of inner coefficients multi-scale product and multi-scale sum are two methods for obtaining spectrum sensing with continuous wavelets. Taking the multi-scale and wavelet transforms, and then estimating the edges given in [29, 32], the multi-scale product technique includes determining discontinuities in a signal's PSD:

$$p(f) = \prod_{j=1}^{J} W's = 2jSr(f) \tag{4}$$

where j is the scaling function's upper limit; $S_r(f)$ is the received signal's PSD; $W'_s = 2^j$ is the first-order derivative at scale $s = 2^j$, and $p(f)$ is the multiscale product.

The spectral boundaries are assumed to be represented by discontinuities in the PSD in this calculation. The energy is calculated for each sub-band to get the spectrum occupancy.

The multi-scale sum technique [33] is based on the idea that various signals at different scales have varying cross scales of information. This means that at different scales, wavelet transformations convey information about the Lipschitz exponent at acute variation spots. As a result, the multi-scale sum technique is employed to preserve signal information at all scales while avoiding attenuation. The multi-scale sum at the *j*th dyadic scale for a CWT is provided as:

$$XjSr(f) = \sum_{j=1}^{J} W2jSr(f) \tag{5}$$

The following are the benefits and drawbacks of the continuous wavelet transform-based spectrum sensing technique:

Advantages

i. CWT allows for enhanced transient localization and oscillatory behavior characterization in a signal.

ii. Because of its fine-grained resolution, CWT is frequently used for singularity detection.

iii. Because of its narrow sample scales, CWT has excellent fidelity in signal analysis.

Disadvantages

i. Because CWT contains a lot of redundancy and is computationally demanding, it's frequently utilized for offline analysis.

ii. For an evaluated signal, CWT does not offer phase information.

iii. A perfect reconstruction of an original signal from CWT coefficients is impossible.

5.2 Discrete Wavelet Transform (DWT) based spectrum sensing

The discrete wavelet transform (DWT) breaks down an input signal $x[m]$ into coarse and fine information. The decomposition, which allows the DWT to examine a signal at multiple frequency bands and resolution, is accomplished by filtering the time-frequency domain signal with successive high pass and lowpass filters [34, 35]. This can be stated mathematically as (**Figure 3**)

$$y_{\text{low}}[k] = \sum_m x[m]h_0[2k - m] \tag{6}$$

$$ylow[k] = \sum_m x[m]h0[2k - m] \tag{7}$$

$y_{\text{high}}[k]$ is the high pass filter output and $y_{\text{low}}[k]$ is the lowpass filter output. Mathematically, the scaling (c_j) and wavelet (d_j) coefficients are represented in [36, 37] as:

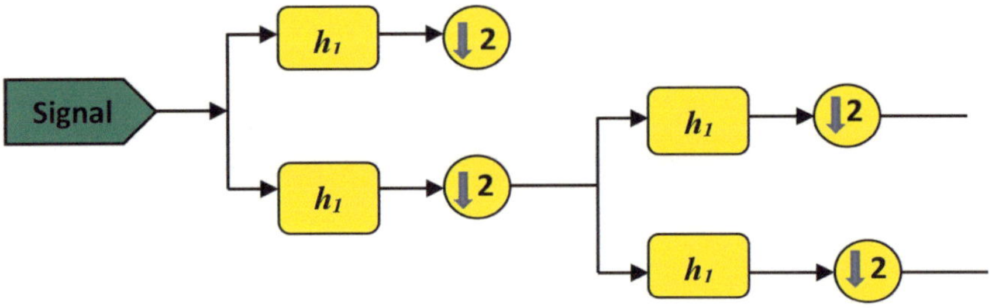

Figure 3.
DWT Filter Bank.

$$c_j(k) = \sum_m h_0(m - 2k)c_{j+1}(m) \tag{8}$$

$$dj(k) = \sum_m h1(m - 2k)cj + 1(m) \tag{9}$$

Advantages and disadvantages of Discrete wavelet transform-based spectrum sensing technique:

Advantages

i. By using a subset of coefficients to capture significant aspects of signals, DWT enables the sparse representation of many natural signals.

ii. DWT achieves signal compression by providing a high-quality approximation of a signal. This is accomplished by deleting several of its close-to-zero coefficients.

iii. Because they have nonredundant orthonormal bases, DWT has flawless reconstruction and is computationally efficient.

Disadvantages

i. Shift sensitivity is a problem with DWT, in which a change in the input signal induces an unanticipated change in the transform coefficients.

ii. The DWT representation of signals in image processing suffers from inadequate directionality, which impairs its optimality.

iii. DWT lacks phase information, which is critical for describing a function's amplitude and local behavior.

5.3 Discrete wavelet packet transform (DWPT) based spectrum sensing

The discrete wavelet packet transform (DWPT) works similarly to the discrete wavelet transform, with the exception that the DWPT transform decomposes both the approximation and detail spaces of a signal [38, 39]. The structure of the DWPT is shown in **Figure 4** [40]. As shown in **Figure 4**, DWPT decomposes a signal $x[n]$ into 2 L sub-bands, where L is the decomposition level.

To perform spectrum sensing, the energy in each sub-band is calculated [40] and compared to a threshold to determine if the sub-band is occupied or not by the principal user. In Ref. [41, 42] gives the following formula for calculating the energy in each sub-band:

$$E = \frac{1}{T}\int_0^T \left[\sum_{j \geq j0} \sum_k cj, k\phi j, k(t) + dj, k\psi j.k(t) \right] 2dt \tag{10}$$

$$E = \frac{1}{T} \sum_{j \geq j0} \sum_k (cj2, k + d2j, k) \tag{11}$$

Advantages and disadvantages of Discrete wavelet Packet transform-based spectrum sensing technique.

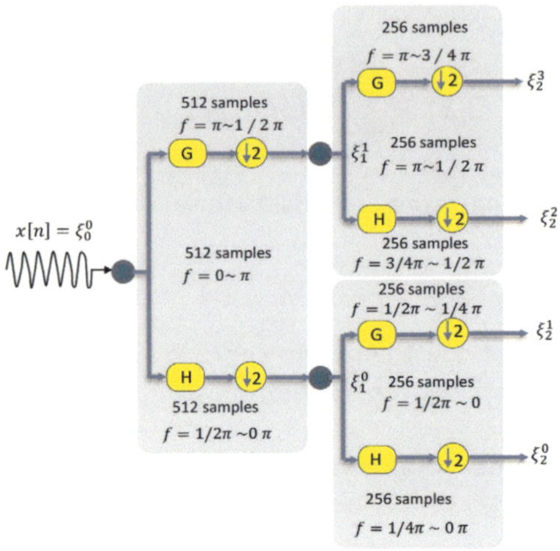

Figure 4.
DWT Structure.

Advantages

i. Because it decomposes both the high and low-frequency components of an input signal, DWPT has more fidelity than DWT.

ii. Without assuming any statistical attribute of the signal, DWPT exhibits good universality in adapting its transform to a signal.

Disadvantages

i. In a DWPT, the high-pass coefficients oscillate around signal singularities.

ii. The nonlinear frequency perception phenomena are ignored by DWPT in voice recognition.

Advantages and disadvantages of wavelet-based spectrum sensing technique. Wavelets, like any other spectrum sensing technology, have advantages and disadvantages. However, when it comes to dynamic frequency management, the benefits of wavelet sensing techniques exceed the drawbacks. The following are some of the benefits of wavelet-based spectrum sensing:

i. Wavelet-based methods are used to accomplish significant data compression. As a result of this compression, the number of sensing measures required is reduced, enhancing estimate speed and lowering the communication power required for transmission. In terms of power source longevity, a reduction in necessary transmission power is a significant benefit for mobile communication devices.

ii. Wavelet-based estimates for sharp-featured sources have the advantage of having fewer side-lobes and thus fewer leakages than most older methods.

DOI: http://dx.doi.org/10.5772/intechopen.102914

iii. Wavelets have a strong ability to modify the time-frequency window to retain orthogonality, allowing them to identify dynamic changes in statistical parameters in any spectrum.

iv. The effect of noise and interference on the signal can be reduced by using a variable time-frequency window design.

v. The tuning capability and flexibility of wavelet bases are used to mitigate channel problems such as inter-symbol interference (ISI) and inter-carrier interference (ICI).

Wavelets do not require a cyclic prefix or guard bands in OFDM applications, making them more efficient than the Fourier transform in terms of spectrum consumption.

The following are some of the drawbacks of wavelet-based spectrum sensing:

i. Frequency selection is not natural.

ii. Requires modification to carry signal phase information.

iii. Requires the creation of filters that implement the required wavelet to meet very strict criteria.

iv. The higher the level of decomposition in the discrete wavelet transform and discrete wavelet packet transform, the more complicated the system becomes.

6. Conclusions

A number of traditional spectrum sensing methods have been studied in this chapter, and their disadvantages have been highlighted. The use of Wavelets' unique properties in Spectrum Sensing has also been explained. A Wavelet Theory of Edge Detection analysis has also been presented. Wavelet categorization has also been briefly addressed. Finally, wavelet-based spectrum sensing was explored, as well as numerous ways to apply the wavelet technique to spectrum sensing, as well as their benefits and drawbacks. Wavelets-based spectrum sensing approach merits, short-comings, and applications have also been examined.

Acknowledgements

Replace the entirety of this text with acknowledgments. Usually, the acknowledgments section includes the names of people or institutions who in some way contributed to the work, but do not fit the criteria to be listed as the authors.

The authorship criteria are listed in our Authorship Policy: https://www.intechopen.com/page/authorship-policy.

This section of your manuscript may also include funding information.

Conflict of interest

If you have any conflicts of interest, please declare them here. If no conflict exists, please put the text: "The authors declare no conflict of interest." or delete this entire section.

Notes/thanks/other declarations

Place any other declarations, such as "Notes", "Thanks", etc. in before the References section. Assign the appropriate heading. Do NOT put your short biography in this section. It will be removed.

Appendices and nomenclature

Assume a frequency band of interest, as illustrated in **Figure 2**, that is further subdivided into N sub-bands, each having its frequency bounds at $f0 < fn$. The objective is to find any empty areas between two neighboring sub-bands so that a secondary user can use them without interfering with the primary. In the absence of noise, the normalized power spectral density (PSD) in each sub-band is:

$$\int_{Fi-1}^{Fi} Si(f)df = Fi - Fi - 1 \tag{A1}$$

Power Spectral density of observed signal $r(t)$:

$$Sr(f) = \sum_{n=1}^{N} \alpha 2nSn(f) + S\omega(f), f \in [f0, fN] \tag{A2}$$

Where $\alpha 2n$ indicates the signal power density within the nth band and $S_n(f)$ is PSD of each sub-band. PSD inside each sub-band is estimated as:

$$= \frac{1}{Fi - Fi - 1} \int_{Fi-1}^{Fi} Sr(f)df \tag{A3}$$

The wavelet transform properties are used to identify the frequency boundaries between successive sub-bands. The following is a one-scale dilation of the wavelet functions ψs:

$$\psi s(f) = \frac{1}{s}\psi\left(\frac{f}{s}\right) \tag{A4}$$

The continuous wavelet transform is defined as:

$$WsSr(f) = Sr\psi s(f) \tag{A5}$$

It gives localized information about $S_r(f)$ at fine scales, and we must identify its first and second derivatives to find abnormalities in it. The first derivative is as follows:

$$Ws1Sr(f) = s\frac{d}{df}(Sr\psi s)(f) \tag{A6}$$

The second derivative is:

$$Ws2Sr(f) = s2\frac{d2}{df2}(Sr\psi s)(f) \tag{A7}$$

Different scales propagate edges and discontinuities. With dyadic scales s = 2j −2, 1–1, 2, 3 ... j, the CWT is obtained. The multi-scale product of the J CWT gradients is used to follow the propagation of edges and discontinuities over multiple scales:

$$p(f) = \prod_{j=1}^{J} W's = 2jSr(f) \tag{A8}$$

Author details

Himanshu Monga[1,2*], Dikshant Gautam[1,2] and Saksham Katwal[1,2]

1 Department of Electronics and Communication Engineering, Jawaharlal Nehru Government Engineering College, Sunder Nagar, Himachal Pradesh, India

2 Jawahar Lal Nehru Government Engineering College, Directorate of Technical Education, Sunder Nagar, Mandi, Himachal Pradesh, India

*Address all correspondence to: himanshumonga@gmail.com

References

[1] Mallat S, Zhong S. Characterization of signals from multiscale edges. Pattern Analysis and Machine Intelligence, IEEE Transactions. 1992;**14**(7):710-732

[2] Adoum BA, Jeoti V. Cyclostationary feature based multiresolution spectrum sensing approach for DVB-T and wireless microphone signals. In: International Conference on Computer and Communication Engineering. Piscataway, New Jersey, United States: IEEE; 2010. pp. 1-6

[3] Kim K, Akbar IA, Bae KK, Um J, Spooner CM, Reed JH. Cyclostationary approaches to signal detection and classification in cognitive radio. In: Proceedings of IEEE Dynamic Spectrum Access Nets. Piscataway, New Jersey, United States: IEEE; 2007. pp. 212-215

[4] Alexender M, Wyglinski M, Nekovee M, Hou T. Cognitive Radio Communications and Networks. USA: Academic Press; 2010

[5] Liang YC, Zeng Y, Peh EC, Hoang AT. Sensing-throughput tradeoff for cognitive radio networks. IEEE Transaction on Wireless Communications. 2008;7(4):1326-1337

[6] Raghave S, Saravanan R, Muthaiah R. Wavelet and S-transform based spectrum sensing in cognitive radio. International Journal of Engineering and Technology (IJET). 2013;**5**(1):147-152

[7] Proakis JG, Manolakis DG. Digital Signal Processing: Principles, Algorithms, and Applications. 4th ed. Prentice Hall, Inc: Hoboken, New Jersey, United States; 2007

[8] Porat B. Course in Digital Signal Processing. New York: John Wiley and Sons, Inc; 1996

[9] Cabric D, Mishra SM, Brodersen RW. Implementation issues in spectrum sensing for cognitive radios. In: Proceedings of the Asilomar Conference on Signals and Systems. Piscataway, New Jersey, United States: IEEE; 2004

[10] Gardner WA. The spectral correlation theory of cyclostationary time-series. Signal Processing. 1986;**11**:13-36

[11] Thomson DJ. Spectrum estimation and harmonic analysis. Proceedings of IEEE. 1982;**70**(9):1055-1096

[12] Narendar M, Vinod AP, Madhukumar AS, Krishna AK. An algorithm for spectrum sensing in cognitive radio using tree-structured filter bank. In: 17th International Conference on Telecommunications. Piscataway, New Jersey, United States: IEEE; 2010

[13] Lakshmanan MK, Ariananda DD, Nikookar H. Cognitive radio transmission and spectrum sensing using a wavelet packet transceiver. In: IEEE 20th International Symposium on Personal, Indoor and Mobile Radio Communication. Piscataway, New Jersey, United States: IEEE; 2009

[14] Akpar S, Francis A. Towards an ICU clinical decision support system using data wavelets. In: International Conference on Intelligent Systems and Data Processing. Piscataway, New Jersey, United States: IEEE; 2011. pp. 37-43

[15] Lee F. Conceptual Wavelets in Digital Signal Processing. California, United States: Space and Signals Technologies; 2009;**6**

[16] Zafar G, Thakare V. Wavelet based feature extraction technique for face

recognition and retrieval: A review. IOSR Journal of Computer Engineering. 2016:49-54

[17] Prasad PM, Prasad DY, Rao S. Performance analysis of orthogonal and biorthogonal wavelets for edge detection of X-ray Images. In: 4th International Conference on Recent Trends in Computer Science and Engineering. Vol. 87. Amsterdam, Netherlands: Elsevier; 2016. pp. 116-121

[18] Carlo C. Shannon wavelets theory. Mathematical Problems in Engineering. 2008;**2008**:1-24. DOI: 10.1155/2008/164808

[19] Walker SJ. A Primer on Wavelets and their Scientific Applications. 2nd ed. Boca Raton: CRC Press; 1999

[20] Teena V, Vidya C, Dipti P. The haar wavelet and biorthogonal wavelet transforms of an image. International Journal of Engineering Research and Applications. 2012;**3**:288-291

[21] Lucas M, Gregory B. Compactly supported wavelets based on almost interpolating and nearly linear phase filters (coiflets). Applied and Computational Harmonic Analysis. 1999;**7**:184-210

[22] Dremin IM, Ivanov V, Nechitailo VA. Wavelets and Their Use. 2001. Available from: https://arxiv.org/pdf/hep-ph/0101182.pdf [Accessed: 16 December 2014]

[23] Guo L, Jun C. A wavelet-based approach for generating individual jumping loads. In: Proceedings of the 34th IMAC, A Conference and Exposition on Dynamics of Multiphysical Systems. Piscataway, New Jersey, United States: IEEE; 2016. pp. 181-194

[24] Amy NR. Investigation of the morlet wavelet for non-linearity detection. In: 2002 IMAC–XX Conference & Conference on Structural Dynamics. Piscataway, New Jersey, United States: IEEE; 2002. pp. 1370-1375

[25] Hugo L, Pascal S, Aude C. New application of wavelets in magnetotelluric data processing: Reducing impedance bias. Earth, Planets and Space. 2016;**68**(70):1-11. DOI: 10.1186/s40623-016-0446-9

[26] Agnieszka R, Robert JB, Pawel M. Wavelet characteristics of hydrological and dissolved oxygen time series in a lowland river. Acta Geophysica. 2016; **64**(3):649-669. DOI: 10.1515/acgeo-2016-0023

[27] Chao Z, Hong L, Qiyun J, Min X. QRS complex detection using combination of Mexican hat wavelet and complex morlet wavelet. Journal of Computers. 2013;**8**(11):2951-2958

[28] Rehman A, Gereth GS, Stephen D. Denoising electromyographic signals via stationary wavelet decomposition and filtering. The Tower. 2014;**6**(2):120-129

[29] Zhi T, Georgios B. A wavelet approach to wideband spectrum sensing for cognitive radios. In: Proceedings of the 1st International Conference on Cognitive Radio Oriented Wireless Networks and Communications. Piscataway, New Jersey, United States: IEEE; 2006. pp. 1-5

[30] Shiann-Shiun J, Jia-Ming C, Hong-Zong L, Chen-Wan T. Wavelet-based spectrum sensing for cognitive radios using hilbert transform. International Journal of Electrical, Computer, Energetic, Electronic and Communication Engineering. 2011;**5**(3): 278-282

[31] Monga H, Rani S, Mehta P. Performance Evaluation of Wavelet-

Based Optical Wireless System using On-Off Keying Modulation. Journal of Optical Communications. 2019;**40**(3): 261-264. DOI: 10.1515/joc-2017-0051

[32] Almedia EP, Carvalho PH, Viera RD. Experimental study of a wavelet-based spectrum sensing techniques. In: Proceeding of the 42nd Asicomar Conference on Signal Systems and Computers. Piscataway, New Jersey, United States: IEEE; 2008. pp. 1552-1556

[33] Karthik D, Naveen MP, Shri H. Wavelet-based spectrum sensing techniques for cognitive radio–a survey. International Journal of Computer Science and Information Technologies. 2011;**3**(2):123-137

[34] Varadharajan E, Rajkumari M. Discrete wavelet transform based spectrum sensing in cognitive radios using eigen filter. International Journal of Advanced Engineering and Technology. 2012;**3**(1):1-3

[35] Chin-Fa H, Tsung-Han T, Chih-Hung L, Shu-Chung Y. An efficient architecture of 1D discrete wavelet transform for noise reduction. International Journal of Advanced Computer Technology. 2013;**5**(3): 412-419

[36] Sidney B, Ramesh G. Introduction to Wavelets and Wavelet Transforms. New Jersey, USA: Prentice-Hall; 1998

[37] Vinit G, Abhay K. Wavelet based dynamic spectrum sensing for cognitive radio under noisy environment. In: International Conference on Modelling Optimization and Computing. Amsterdam, Netherlands: Elsevier; 2012. pp. 3228-3234

[38] Shrutika S, Kumbhar S. Wavelet packet transform based energy detector for spectrum sensing. International Journal of Emerging Technology and Advanced Engineering. 2012;**2**(3): 379-382

[39] Haleh H, Sharifah K, Norsheila F, Ali F. Compressed wavelet packet based spectrum sensing with adaptive thresholding in cognitive radio. Canadian Journal of Electrical and Computer Engineering. 2015;**38**(1):31-36

[40] Venkat R, Nikookar H. Performance evaluation of a wavelet packet based spectrum estimator for cognitive radio applications. In: 18th IEEE Symposium on Communications and Vehicular Technology. Piscataway, New Jersey, United States: IEEE; 2011. pp. 1-6

[41] Zhinjin Q, Wang N, Yue G, Laurie C. Adaptive threshold for energy detector based on discrete wavelet packet transform. In: Wireless Telecommunications Symposium. Piscataway, New Jersey, United States: IEEE; 2012. pp. 1-5

[42] Wei N, Chen Z, Zhu A. Research on adaptive resolution spectrum sensing method based on discrete wavelet packet transform. Telkomnika Indonesian Journal of Electrical Engineering. 2014; **12**(2):1385-1394

Chapter 6

Monitoring the Condition of Railway Tracks Using a Convolutional Neural Network

Hitoshi Tsunashima and Masashi Takikawa

Abstract

Condition monitoring of railway tracks is effective for the sake of an increase in the safety of regional railways. This study proposes a new method for automatically classifying the type and degradation level of track fault using a convolutional neural network (CNN), which is a machine learning method, by imaging car body acceleration on a time-frequency plane by continuous wavelet transform. As a result of applying this method to the data measured in regional railways, it was possible to classify and extract the sections that need repair according to the degree of deterioration of the tracks, and to identify the track fault in those sections.

Keywords: railway, track, condition monitoring, wavelet, convolutional neural network

1. Introduction

Maintenance of railway tracks is essential for the safe operation of trains. Railway operators conduct track inspections using track geometry cars and track maintenance crews. However, regional railway operators, who carry fewer passengers, often lack the personnel and funds to conduct adequate track inspections. The monitoring of railway track geometry from an in-service vehicle has become increasingly attractive over the past decade [1].

To address this problem, a system that can monitor the track condition inexpensively and frequently using a device incorporating sensors and a global navigation satellite system (GNSS) unit, which is installed on in-service trains, has been developed [2, 3]. The system calculates root mean square (RMS) values from the vertical acceleration, lateral acceleration, and roll angular velocity of the car body. To select sites for repair, we adopt the method of prioritizing sites with the highest numerical values.

The acceleration RMS is closely related to the general health of the track [4]. In Ref. [5], RMS values are used to identify track irregularities for longitudinal level, alignment, cross-level. However, monitoring based on RMS values alone is not sufficient. Without frequency information, it is difficult to identify the type of track fault. Furthermore, since the amount of data generated by constant measurement is enormous, it is necessary to automate the analysis in order to monitor and predict the track condition efficiently.

In this study, we propose a method to classify the types of track faults automatically by means of machine learning, using a CNN trained on images created via a CWT from the vibration acceleration on the time-frequency plane. A continuous wavelet transform (CWT) is a transformation technique that emphasizes certain portions of the waveform by suppressing other portions as it proceeds by multiplying a target waveform using a mother wavelet [6]. A convolutional neural network (CNN) is a class of deep neural networks. It is widely used for image recognition.

To verify the effectiveness of the algorithm we developed, we first describe the results of simulating the vibration of a car body when passing over a faulty track. Next, we describe the results of diagnosing track faults from the vertical vibration acceleration data of a car body measured by a regional railway.

2. Literature review of track condition monitoring using machine learning techniques

It should be necessary for railway operators to control track irregularity, such as vertical rail profiles, lateral alignment, gauge, cross-level, twist (depicted in **Figure 1**) properly. Track irregularities cause vehicle vibrations that degrade the rider's comfort and increase the risk of derailments. Track irregularities are strongly correlated with vehicle vibrations. Thus, it can be possible to estimate general trends of the track condition by analyzing vehicle vibrations.

Although track geometry measurement systems using in-service vehicles are becoming increasingly attractive around the world [2, 7–9], the repeated checking of the same track provides the information regarding track geometry degradation, which can be fed back to the track maintenance section for taking essential actions. The use of vehicle responses in the track geometry assessment process allows identifying of

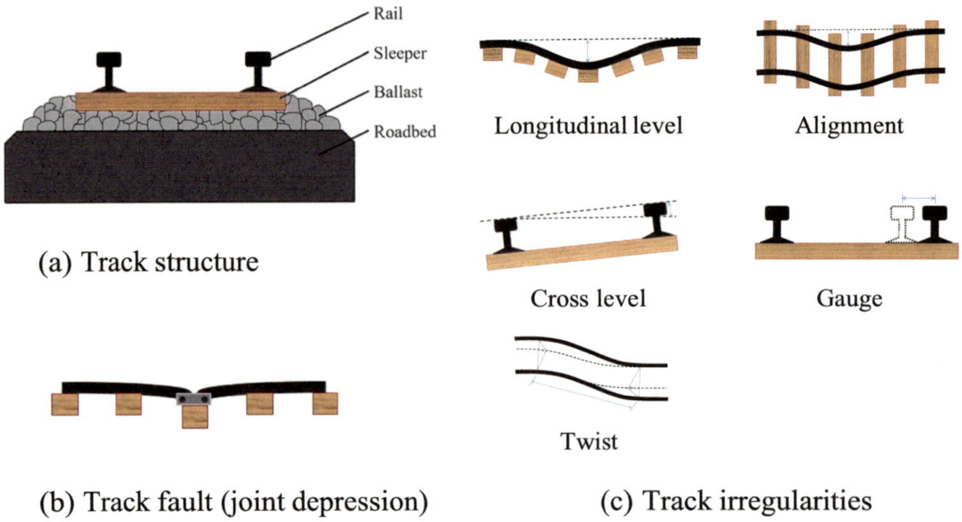

Figure 1.
Track structure and irregularities.

critical defects, which could not have been identified from geometry parameters, and thus, improve the maintenance operations.

Tsunashima et al. proposed techniques of condition monitoring of railway tracks based on time-frequency analysis [10]. They compared the performance of Hilbert-Huang transforms (HHT) and CWT for identifying track faults from car body vibration. It is shown that the feature of track fault can be identified in time-frequency plane.

Tsunashima proposed a classifier based on a machine learning technique for identifying track faults automatically from measured car body vibration [5]. It is shown that the degradation of track can be classified in the feature space consisting of car body vibration RMS.

Faghih-Roohi et al. proposed a deep convolutional neural network for the analysis of image data for the detection of rail surface defects [11]. They explored the efficiency of the proposed deep convolutional neural network for the detection and classification of rail surface defects.

Zheng et al. proposed a multi-object detection method based on a deep convolutional neural network that can achieve non-destructive detection of rail surface and fastener defects [12]. A defect detection model based on Mask R-CNN and ResNet framework was utilized to detect the surface defects.

Jin et al. proposed a machine learning framework based on wavelet scattering networks and neural networks for identifying railhead defects [13].

Alvarenga *et al.* proposed an embedded system for online detection and location of rails defects based on eddy current [14]. They proposed a new method to interpret eddy current signals by analyzing their wavelet transforms through a convolutional neural network.

3. Effect of track faults on time-frequency plane

3.1 Overview of the simulation

When a train runs on a track, vibrations that correspond to the track geometry are generated [15, 16]. Therefore, in this study, to verify the relationship between the type of track fault and the car body vibration acceleration, and to evaluate the effectiveness of time-frequency analysis in detecting track faults, we simulated the occurrence of track faults, calculated the vertical vibration acceleration of the car body, and then applied a CWT, a method of time-frequency analysis, to the results.

3.2 Continuous wavelet transform (CWT)

A CWT is a method that simultaneously detects the frequency and time characteristics of an unsteady signal, by comparing the original signal with dilated and translated versions of a small wavelike function called the mother wavelet. Using this method, it is possible to view the amplitude and frequency information of the vibration acceleration as an image. In this study, we used the *Morlet* wavelet, which offers a relatively good balance between localization of time and frequency, as the mother wavelet [17] (see Appendix A).

This technique is well suited for analyzing unsteady signals, such as $x(t)$ those that exhibit sudden variation, and is defined as follows:

$$W_\psi(a,b) = \int_{-\infty}^{\infty} \frac{1}{\sqrt{a}} \psi^* \left(\frac{t-b}{a}\right) x(t) dt, \tag{1}$$

where, variables a and b correspond to the dilatation and location parameters, respectively, they translate the mother wavelet $\psi(t)$ by a time shift b in time, and by $1/a$ in frequency. indicates the complex conjugate of ψ.

3.3 Vehicle model used in the simulation

The vehicle model used in the simulation is shown in **Figure 2** [10]. The vehicle model consists of a total of seven rigid bodies: one car body, two bogies, and four wheelsets. The car body and bogie were assigned two degrees of freedom (DOF) for bounce and pitch, and the wheelset was assigned one DOF for the bounce. The vehicle's parameters were obtained from measurement data from a regional railway vehicle equipped with an onboard sensing device.

3.4 Simulation conditions

In the simulation, the vehicle model was run at 60 [km/h] for 500 [m], and the results were output for the section between 100 [m] and 350 [m]. We set rail joint faults (joint depressions) at 4 points; otherwise, the track was assumed to be straight. To set the rail joint faults, we used the function model shown in **Figure 3** [18].

The geometry of the modeled track are represented by

$$y = Ae^{-\left(\frac{1}{2}\right)\left(\frac{x}{k}\right)^2}, \tag{2}$$

and

$$y_2 = \frac{1}{2}\left(\frac{x}{k}\right)^2 - A. \tag{3}$$

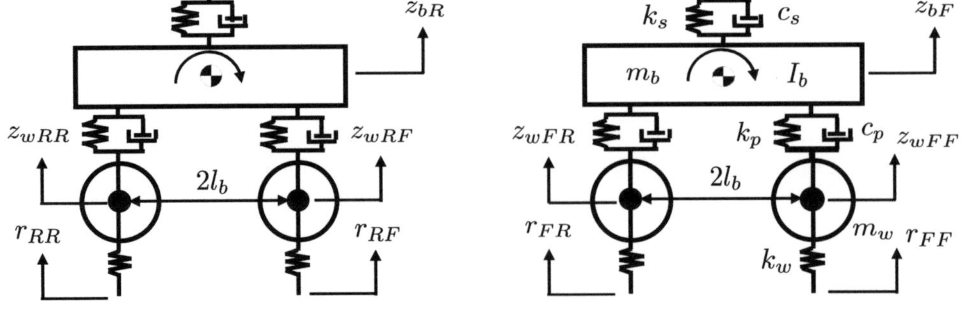

Figure 2.
Vehicle model [10].

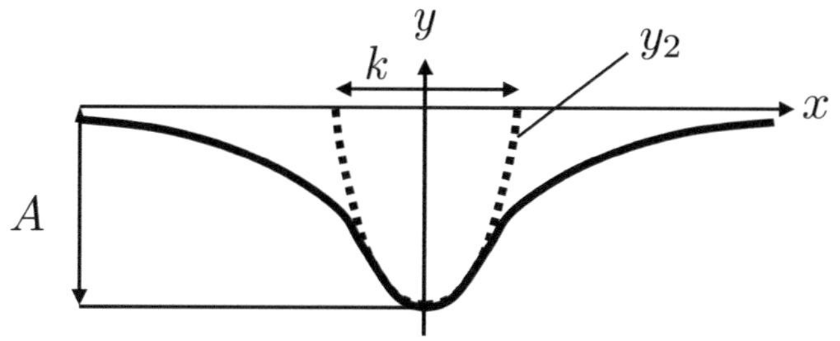

Figure 3.
Track fault model.

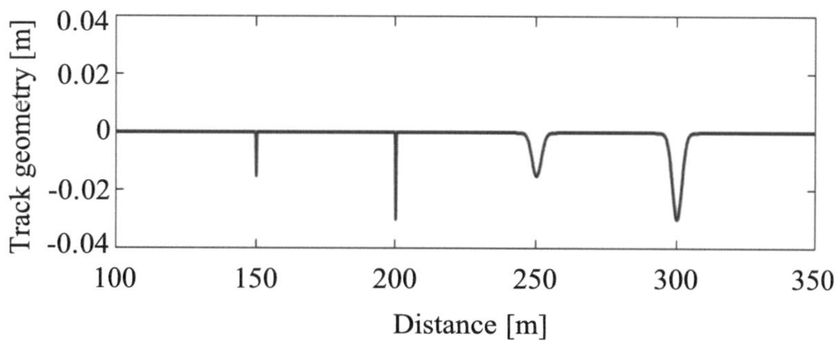

Figure 4.
Track geometry with different faults.

The track geometry used in the simulation is shown in **Figure 4**. At the 150 [m] and 200 [m] points, we set depths of $A = 15$ [mm] and 30 [mm], respectively, to simulate joint depressions, which are depressions in the top surface of the railhead that occur at track joints.

In both cases, we set the depression length $k = 83.5$ [mm]. For comparison, at the 250 [m] and 300 [m] points, we set a gentler dip in track geometry by setting depths of $A = 15$ [mm] and 30 [mm], respectively, but with length $k = 1670$ [mm]. The values of A and k were determined with reference to generally occurring track displacement. The car body vibration acceleration was assumed to occur directly above the center of the front bogie of the vehicle model. The simulation was performed at a sampling frequency of 200 [Hz].

3.5 Simulation results

The simulated vertical vibration acceleration of the car body is shown in **Figure 5a**. The figure shows that characteristic vibrations corresponding to the track geometry are generated at the points where the track faults were set. **Figure 5b** shows the result of the CWT of the simulated vertical vibration acceleration. The color bar indicates the magnitude of the amplitude in the time-frequency plane.

At 150 [m] and 200 [m], the points where the joint depressions were simulated, vibrations in the frequency band of 15–30 [Hz] were detected due to the impulse-like track geometry, and variations depending on depth A can be seen in the CWT images. In addition, at 250 [m] and 300 [m], the points where gentler dips in the track geometry were set, vibrations in the frequency band of 0–5 [Hz] were detected, and variations depending on depth A can be seen in the CWT images. These results

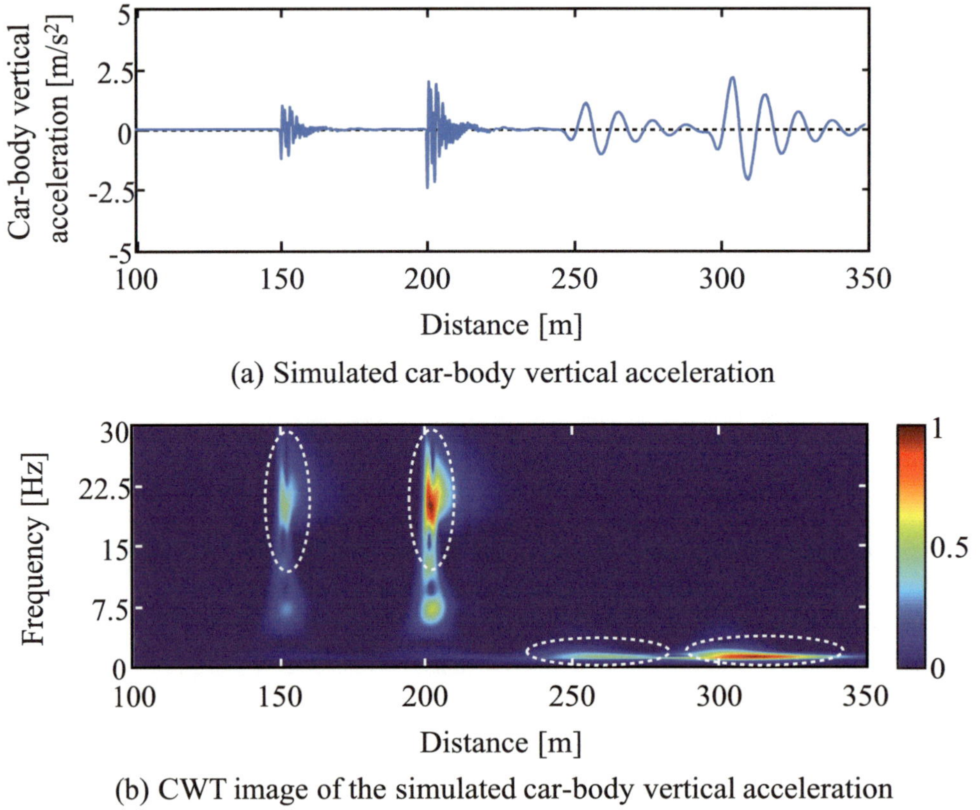

Figure 5.
Simulated car body vertical acceleration and its CWT image.

demonstrate that the CWT images are effective for identifying track faults, since the features of the CWT images change markedly depending on the type and level of degradation.

4. Monitoring the condition of railway track using a convolutional neural network

4.1 Track condition monitoring system

Figure 6 shows the track condition monitoring system developed and applied for regional railway lines in Japan [2].

Accelerometers and rate gyros in the onboard sensing device measure the car body vibration. A GNSS receiver detects the location and speed of the train. Collected data are transmitted to the data server in the monitoring center continuously via a mobile phone network.

The diagnostic software analyses the collected data and results are fed back to the railway operators through online channels via tablet computers. The diagnostic results are used to facilitate the maintenance work of railway operators.

Convolutional neural networks are a method used in the field of machine learning called deep learning and are particularly suitable for image recognition. In this study,

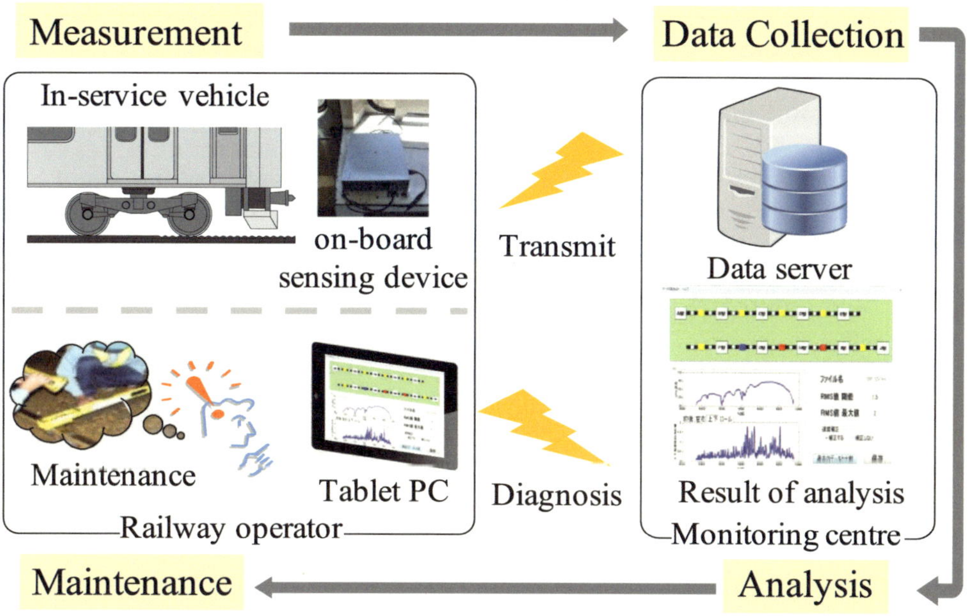

Figure 6.
Track condition monitoring system [2].

we examined the effectiveness of classifying longitudinal level irregularities and joint depressions automatically, using a diagnostic algorithm, we constructed based on a convolutional neural network trained on CWT images generated from vertical vibration acceleration data from a car body. The diagnostic procedure is shown in **Figure 7**.

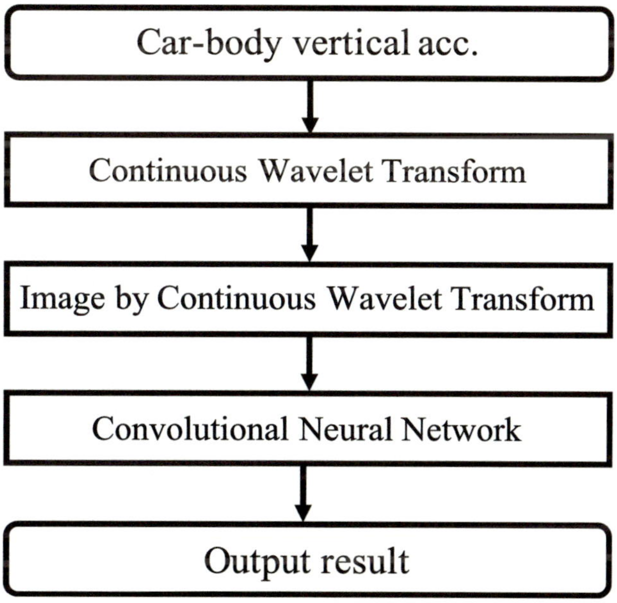

Figure 7.
Diagnostic procedure.

5. Identification of the condition of railway track using vertical vibration acceleration data measured from an actual car body

5.1 Overview of diagnosis

The car body's vertical acceleration with track faults was collected in a regional railway line using the track condition monitoring system. The input data for the classifier consists of vertical vibration acceleration measurements from an onboard sensing device in a car body, which are then converted into images using a CWT. **Figure 8** shows an example of converting the measurements into a CWT image.

The vibration characteristics of the joint depression at the distance of 25.82 [km] appear in the 10–30 [Hz] frequency range. The vibration characteristics of the longitudinal level irregularities around 25.95 [km] appear in the 0–5 [Hz] frequency range.

5.2 Images used for training and evaluation

In this study, we investigated the following three types of diagnoses:

- Classification of images into three types: longitudinal level irregularity, joint depression, and normal.

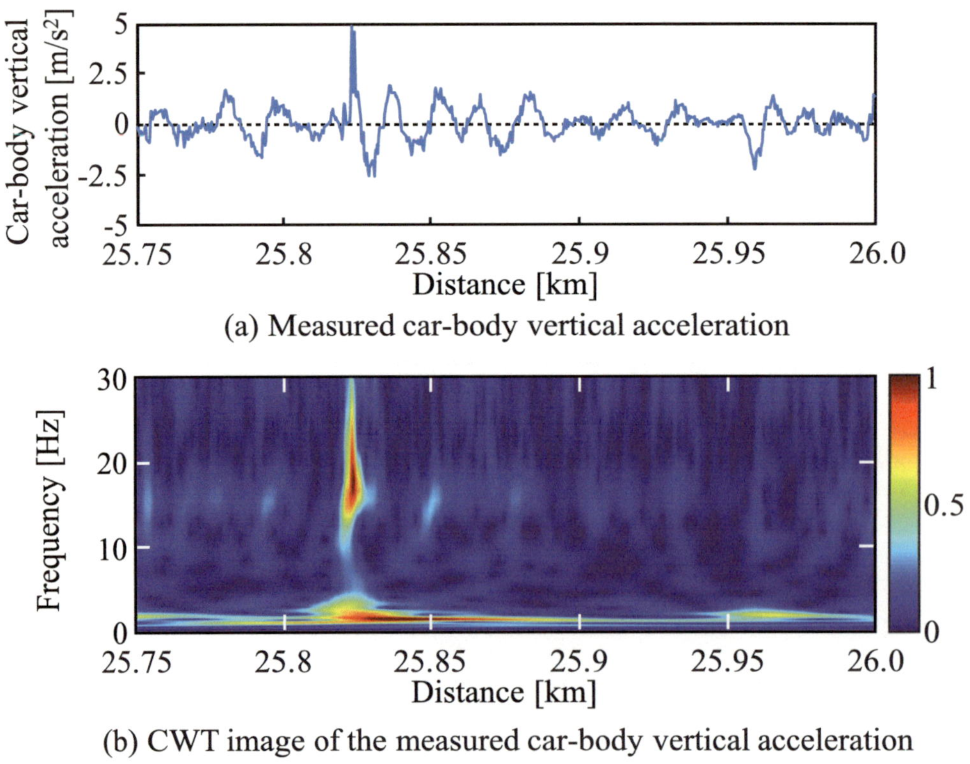

Figure 8.
Measured car-body vertical acceleration and its CWT image.

- Classification of the degradation level of longitudinal level irregularity into normal, medium, and large.
- Classification of the degradation level of joint depression into normal, medium, and large.

Examples of images used for each task are shown in **Figures 9–11**. The images were created with an aspect ratio of 1:1 (150 × 150 pixels), which is optimal for training.

For diagnosing the level of degradation of longitudinal level irregularities, in cases where the one-side amplitude of the vibration acceleration was normal, images of car body acceleration of 0–0.5 $[m/s^2]$ were used. To diagnose medium degradation, images of 0.8–1.2 $[m/s^2]$ were used, and to diagnose large degradation, images of 1.5 $[m/s^2]$ or greater were used.

For diagnosing the level of degradation of joint depressions, in cases where the one-side amplitude of the vibration acceleration was normal, images of body

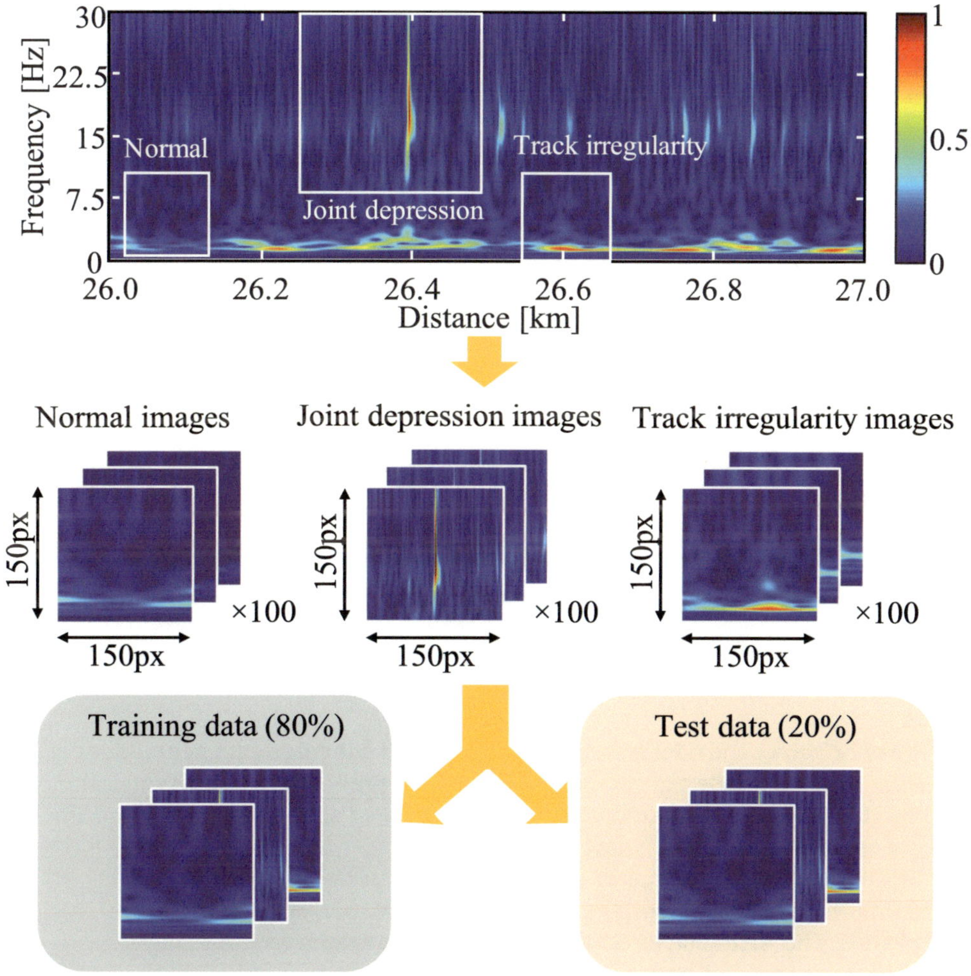

Figure 9.
CWT images of faulty track.

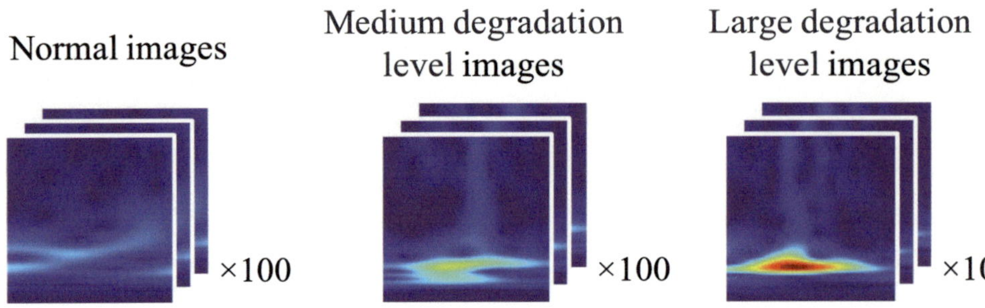

Figure 10.
CWT images of the different levels of a degraded track (track irregularity).

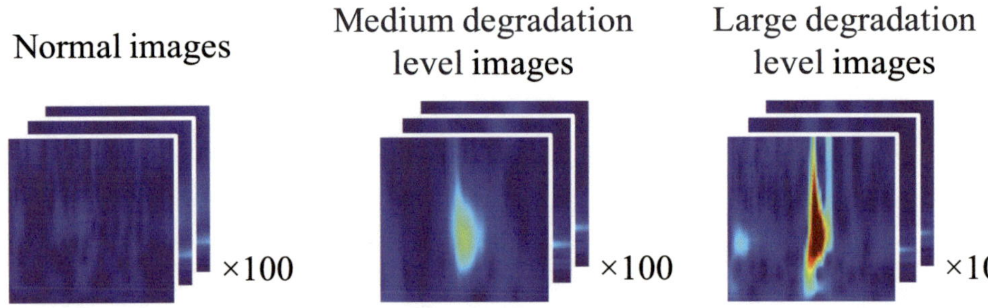

Figure 11.
CWT images of the different levels of a degraded track (joint depression).

acceleration of 0 to were used. To diagnose medium degradation, images of 2.5–3.5 $[\mathrm{m/s^2}]$ were used, and to diagnose large degradation, images of 4.5 $[\mathrm{m/s^2}]$ or greater were used.

5.3 Identification of longitudinal level irregularities and joint depressions

5.3.1 Configuration of the trained convolutional neural network

We prepared a total of 300 images: 100 normal images, 100 images with a longitudinal level irregularity, and 100 images with a joint depression. We set aside 80% of the images for training and 20% for evaluation as shown in **Figure 9**.

Figure 12 shows the configuration of the trained convolutional neural network (see Appendix B). In the figure, the name of the process and the size (vertical × horizontal × channels) before processing are indicated above each layer, and the size after processing is indicated below the layer.

The Convolution layer applies the convolution operation to the image, representing it in matrix form; the Max pooling layer performs information compression; the Affine layer combines information from different layers, and the Output layer outputs a set of probabilities indicating how well the image matches the three types of training image data. The number of training sessions was set to 50.

DOI: http://dx.doi.org/10.5772/intechopen.102672

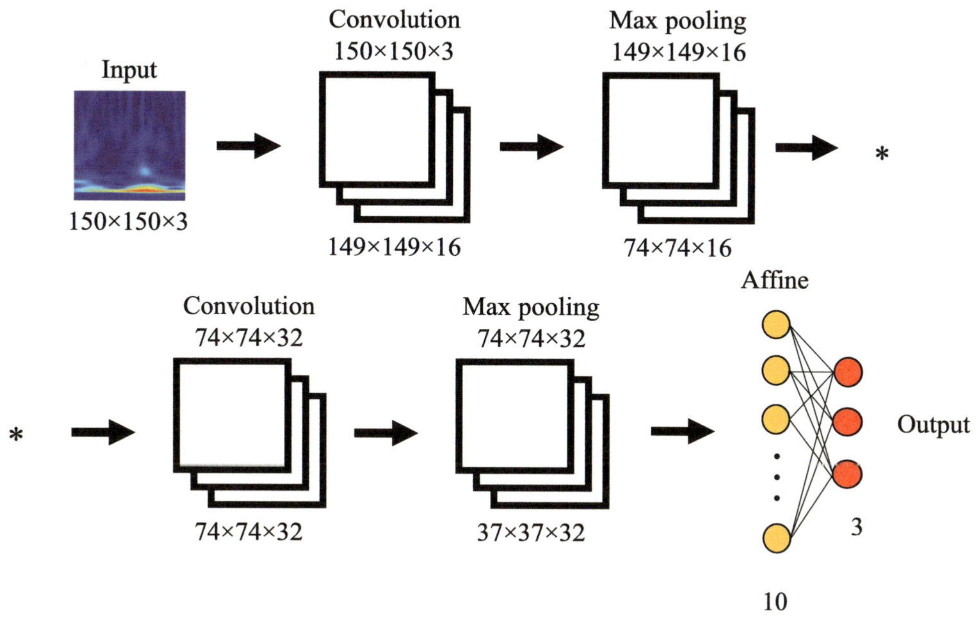

Figure 12.
Network configuration.

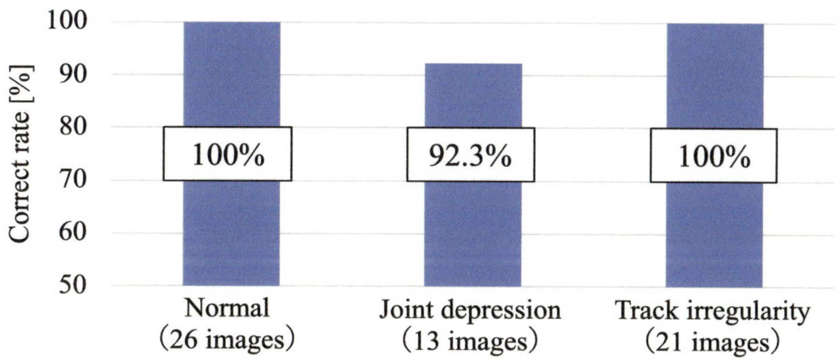

Figure 13.
Detection accuracy for the type of track fault.

5.3.2 Diagnosis results

Figure 13 shows the results of using images for evaluation to discriminate longitudinal level irregularity track faults versus joint depression track faults versus normal track. The overall accuracy rate was 98.3%, demonstrating that convolutional neural networks are effective for the classification of track faults.

5.4 Identification of the degradation level of longitudinal level irregularities

5.4.1 Configuration of the trained convolutional neural network

In order to classify the degradation level of longitudinal level irregularities into three types: normal, medium, and large, we prepared a total of 300 images: 100 normal, 100 medium, and 100 large. We set aside 80% of the images for training and

20% for evaluation. The network configuration and the number of training sessions were the same as in Section 5.3.

5.4.2 Diagnosis results

Detection results using the trained model are shown in **Figure 14**. The overall accuracy rate was 98.3%, demonstrating that the level of longitudinal level irregularity can be classified with high accuracy into normal, medium, and large.

5.5 Classification of the degradation level of joint depression

5.5.1 Configuration of the trained convolutional neural network

In order to classify the degradation level of joint depression into three types: normal, medium, and large, we prepared a total of 300 images: 100 normal, 100 medium, and 100 large. We set aside 80% of the images for training and 20% for evaluation. The network configuration and the number of training sessions were the same as in Section 5.3.

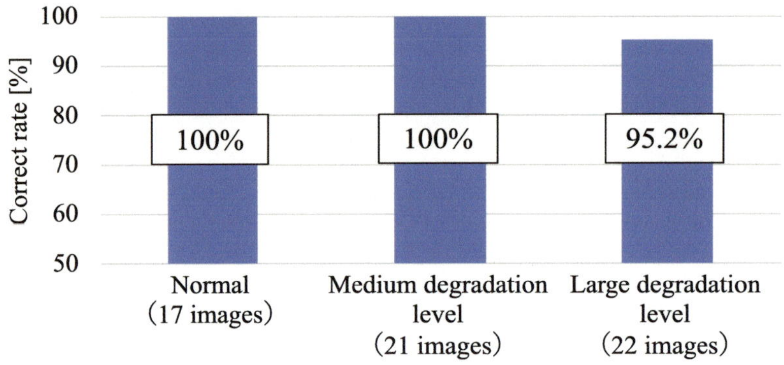

Figure 14.
Detection accuracy for the different levels of a degraded track (track irregularity).

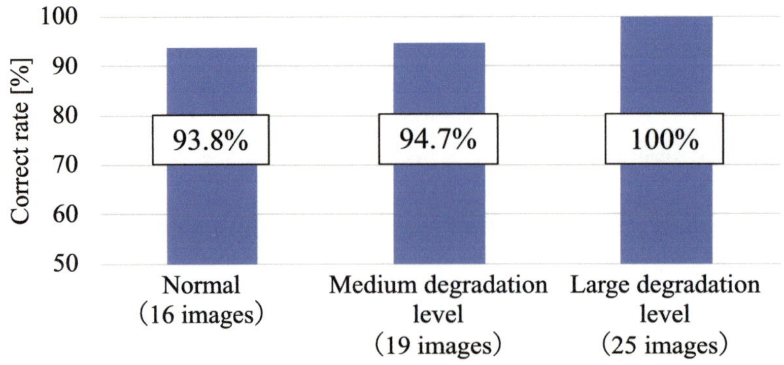

Figure 15.
Detection accuracy for the different levels of a degraded track (joint depression).

5.5.2 Diagnosis results

Detection results using the trained model are shown in **Figure 15**. Some incorrect diagnoses were made in the images of normal and medium joint depression. However, the overall accuracy was 96.7%, which was sufficient to classify the level of joint depression, demonstrating that the diagnostic algorithm we developed is effective for the diagnosis of joint depression.

6. Investigation of CWT images that were diagnosed incorrectly

Figure 16 shows an example of an image that was diagnosed incorrectly. The right side of **Figure 16a** was diagnosed as normal, even though it shows joint depression.

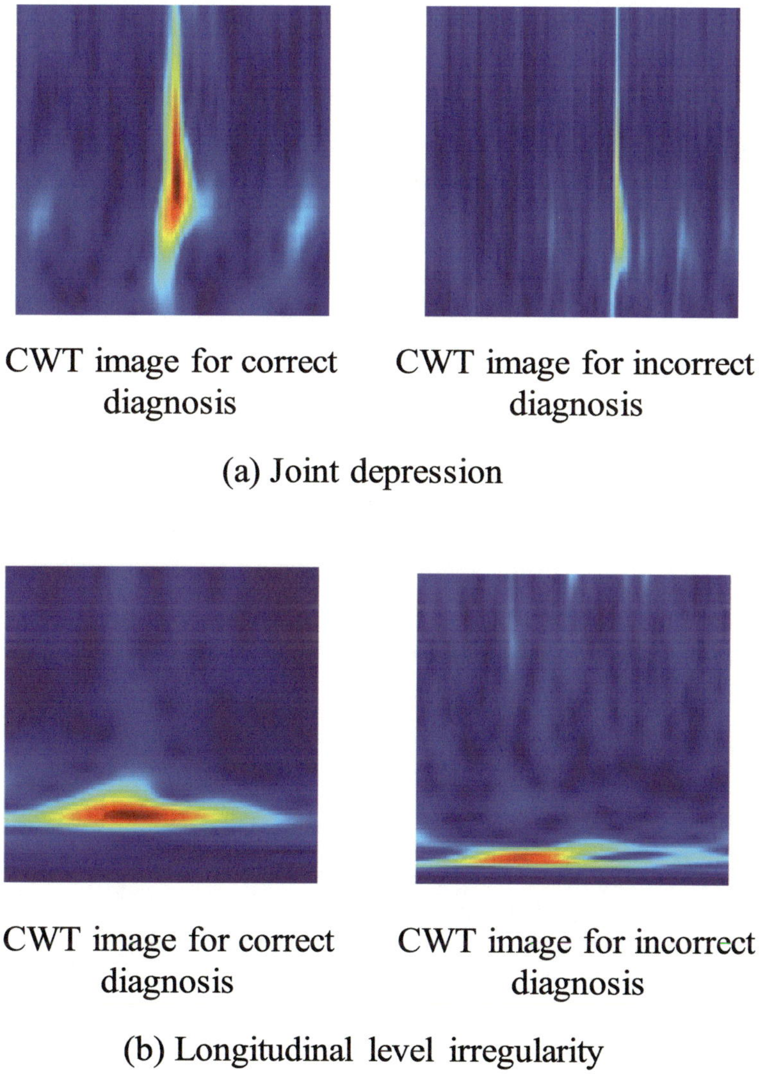

Figure 16.
CWT images that were diagnosed incorrectly.

Conversely, the left side of **Figure 16a** shows an image that was diagnosed correctly as a joint depression. Comparing those, the feature representing the joint depression is extremely small in the incorrectly diagnosed image. This reveals that an incorrect diagnosis can occur when the features are extremely small.

The right side of **Figure 16b** was diagnosed as normal, even though it shows a large track irregularity. Conversely, the left side of **Figure 16b** shows an image that was diagnosed correctly as a large track irregularity. The reason for the incorrect diagnosis was that the large amplitude of the vertical acceleration, shown in red color, was appeared at the bottom of the CWT image.

7. Conclusion

In this study, we proposed a method to classify the type and level of track faults automatically using a convolutional neural network trained on car body vibration acceleration measurements converted into images using a CWT, a well-known method of time-frequency analysis. The algorithm we developed was used to perform the diagnosis of track conditions on actual measurements.

The results demonstrated that it is possible to diagnose the type and level of degradation of track faults with high accuracy.

In future work, we plan to improve the algorithm to estimate the locations of track faults accurately in actual measurements and monitor the condition of railway tracks in more detail.

Acknowledgements

This research was funded by Nihon University Research Grant for Social Implementation (19-006) (2019). We would like to thank Editage (www.editage.jp) for English language editing.

Conflict of interest

The authors declare no conflict of interest.

Abbreviations

CWT	continuous wavelet transform
RMS	root mean square
CNN	convolutional neural network
HHT	Hilbert–Huang transform
GNSS	global navigation satellite system

A. Appendix

A CWT is a method that simultaneously detects the frequency and time characteristics of an unsteady signal, by comparing the original signal with dilated and

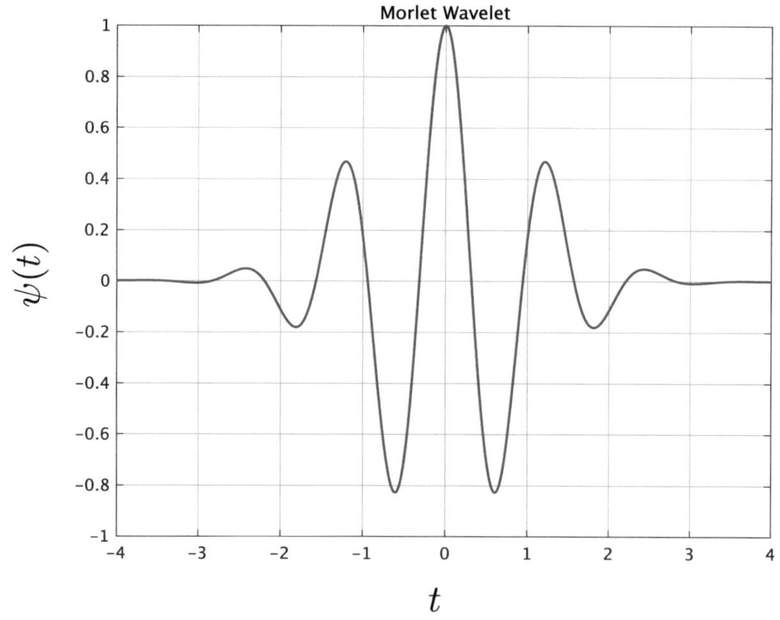

Figure 17.
Real-valued Morlet wavelet.

translated versions of a small wavelike function called the mother wavelet. The CWT computes the inner products of a continuous signal with a set of continuous wavelets according to the following equation

$$W_{\psi}(a,b) = \int_{-\infty}^{\infty} \frac{1}{\sqrt{a}} \psi^* \left(\frac{t-b}{a}\right) x(t)dt, \quad \text{(A1)}$$

where, variables a and b correspond to the dilatation and location parameters, respectively, they translate the mother wavelet $\psi(t)$ by a time shift b in time, and by $1/a$ in frequency. ψ^* indicates the complex conjugate of ψ.

In this study, we used the real-valued *Morlet* wavelet (**Figure 17**) as the mother wavelet $\psi(t)$.

$$\psi(t) = e^{-\frac{t^2}{2}} \cos(5t). \quad \text{(A2)}$$

B. Appendix

A Convolutional Neural Network (CNN) is a well-known deep learning architecture. There are numerous variants of CNN architectures. The basic components of CNN consist of convolutional layer, pooling layer, and fully-connected layers [19].

B.1 Convolution Layer

The objective of the convolution operation is to extract the significant features from the input image. The convolution layer is composed of several convolution kernels which are used to compute different feature maps. The feature maps are generated by the convolution operation with the filter that acts as the feature extractor as follows.

$$I_2(x,y) = \sum_{j=-N}^{N} \sum_{i=-N}^{N} F(i,j) I_1(x-i, y-j). \tag{A3}$$

where $I_1(x,y)$: pixel value of input image at (x,y), $I_2(x,y)$: pixel value of output image at (x,y), $F(i,j)$: filter coefficient.

B.2 Pooling layer

The Pooling layer is responsible for reducing the spatial size of the feature maps. This is to decrease the computational power required to process the data through size reduction. It is useful for extracting dominant features. There are two types of Pooling: Max Pooling and Average Pooling. Max Pooling returns the maximum value from the portion of the image. On the other hand, Average Pooling returns the average value. In this study, Max Pooling were used. **Figure 18** shows the example of the Max Pooling operation.

B.3 Activation function

Rectified linear unit (ReLU) is one of the most famous activation functions. In this study, the following function is used to adjust the output of the Pooling Layer.

$$y = \begin{cases} 0 & (x \leq 0) \\ x & (x > 0) \end{cases} \tag{B1}$$

where x is the input of the activation function. The simple operation of the activation function makes the faster computation than sigmoid or hyperbolic tangent functions.

Softmax function defined by

$$y_i = \frac{e^{x_i}}{\sum_{k=1}^{n} e^{x_k}} \quad (i = 1, 2, \cdots, n), \tag{B2}$$

was used in output layer. Where n indicates the number of classification.

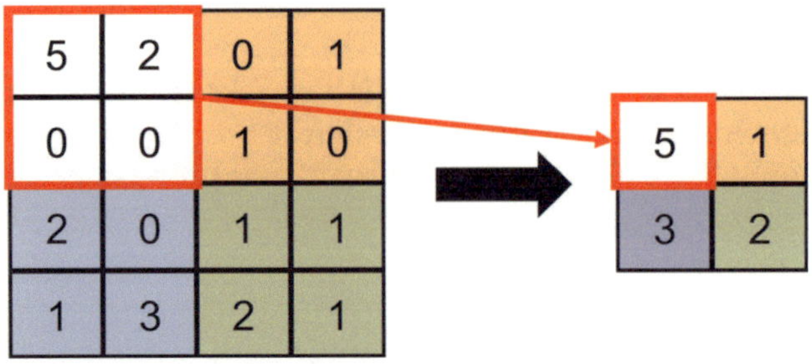

Figure 18.
Max pooling.

B.4 Fully connected layer

In fully connected layers, the neuron applies a linear transformation to the input vector through a weights matrix. In this study, an Affine transformation was used in fully connected layer.

B.5 Loss function

The loss function is the function that computes the distance between the current output of the algorithm and the expected output. In this study, we employed the categorical cross-entropy, which is well suited to classification tasks.

Author details

Hitoshi Tsunashima*† and Masashi Takikawa†
College of Industrial Technology, Nihon University, Chiba, Japan

*Address all correspondence to: tsunashima.hitoshi@nihon-u.ac.jp

† These authors contributed equally.

References

[1] Weston P, Roberts C, Yeo G, Stewar E. Perspectives on railway track geometry condition monitoring from in-service railway vehicles. Vehicle System Dynamics. 2015;**53**(7):1063-1091. DOI: 10.1080/00423114.2015.1034730

[2] Tsunashima H, Mori H, Ogino M, Asano A. Development of track condition monitoring system using onboard sensing device. In: Zboinski K. editors. Railway Research. London: IntechOpen; 2015. p. 145. DOI: 10.5772/61077

[3] Mori H, Ohno H, Tsunashima H, Saito Y. Development of compact size onboard device for condition monitoring of railway tracks. Journal of Mechanical Systems for Transportation and Logistics. 2013;**6**(2):142-149. DOI: 10.1299/jmtl.6.142

[4] Vinberg EM, Martin M, Firdaus AH, Tang Y, Qazizadeh A. Railway Applications of Condition Monitoring. Stockholm, Sweden: KTH Royal Institute of Technology; 2018. p. 147. DOI: 10.13140/RG.2.2.35912.62729

[5] Tsunashima H. Condition monitoring of railway tracks from car-body vibration using a machine learning technique. Applied Sciences. 2019;**9**(13): 2734. DOI: 10.3390/app9132734

[6] Daubechies I. Ten Lectures on Wavelets. Society for Industrial and Applied Mathematics: Philadelphia, Pennsylvania, United States. 1992. p. 441 DOI: 10.1137/1.9781611970104

[7] Weston P, Ling C, Goodman C, Roberts C, Li P, Goodall R. Monitoring vertical track irregularity from in-service railway vehicles. Proceedings of the Institution of Mechanical Engineers, Part F: Journal of Rail and Rapid Transit. 2007;**221**:75-88. DOI: 10.1243/0954409JRRT65

[8] Weston P, Ling C, Goodman C, Roberts C, Li P, Goodall R. Monitoring lateral track irregularity from in-service railway vehicles. Proceedings of the Institution of Mechanical Engineers, Part F: Journal of Rail and Rapid Transit. 2007;**221**: 89-100. DOI: 10.1243/0954409JRRT64

[9] Wei X, Liu F, Jia L. Urban rail track condition monitoring based on in-service vehicle acceleration measurements. Measurement. 2016;**80**:217-228. DOI: 10.1016/j.measurement.2015.11.033

[10] Tsunashima H, Hirose R. Condition monitoring of railway track from car-body vibration using time–frequency analysis. Vehicle System Dynamics. 2020;**1**:1-18. DOI: 10.1080/00423114.2020.1850808

[11] Faghih-Roohi S, Hajizadeh S, Nunez A, Babuska R, De Schutter B. Deep convolutional neural networks for detection of rail surface defects. In: Proceedings of the 2016 International Joint Conference on Neural Networks. Vancouver, Canada: IJCNN; 2016. pp. 2584-2589

[12] Zheng D, Li L, Zheng S, Chai X, Zhao S, Tong Q, et al. A defect detection method for rail surface and fasteners based on deep convolutional neural network. Hindawi, Computational Intelligence and Neuroscience. 2021; **2565500**:15. DOI: 10.1155/2021/2565500

[13] Jin Y. Wavelet scattering and neural networks for railhead defect identification. Materials. 1957;**2021**:14. DOI: 10.3390/ma14081957

[14] Alvarenga TA, Carvalho AL, Honorio LM, Cerqueira AS, Filho LMA,

Nobrega RA. Detection and classification system for rail surface defects based on Eddy current. Sensors. 2021;**21**:7937. DOI: 10.3390/s21237937

[15] Kraft S, Causse J, Coudert F. Vehicle response based track geometry assessment using multi-body simulation. Vehicle System Dynamics. 2018;**56**(2): 190-220. DOI: 10.1080/00423114. 2017.1359418

[16] Karis T, Berg M, Stichel S, Li M, Thomas D, Dirks B. Correlation of track irregularities and vehicle responses based on measured data. Vehicle System Dynamics. 2018;**56**(6):967-981. DOI: 10.1080/00423114.2017.1403634

[17] Le T. Use of the Morlet mother wavelet in the frequency-scale domain decomposition technique for the modal identification of ambient vibration responses. Mechanical Systems and Signal Processing. 2017;**95**:488-505. DOI: 10.1016/j.ymssp.2017.03.045

[18] Garg V, Dukkipati R. Dynamics of Railway Vehicle Systems. Cambridge, Massachusetts: Academic Press; 1984. p. 407

[19] Gu J, Wang Z, Kuen J, Ma L, Shahroudy A, Shuai B, et al. Recent advances in convolutional neural networks. Pattern Recognition. 2018;**77**: 354-377. DOI: 10.1016/j. patcog.2017.10.013

Chapter 7

Image Restoration and Noise Reduction with Context-Dependent Wavelet Graph and ADMM Optimization

Jinn Ho, Shih-Shuo Tung and Wen-Liang Hwang

Abstract

We represent the image noise reduction and restoration problems as context-dependent graphs and propose algorithms to derive the optimal graphs by the alternating direction method of multipliers (ADMM) method. An image is spatially decomposed into smooth regions and singular regions, consisting of edges and textures. The graph representing a smooth region is defined in the image domain, while that representing a singular region is defined in the wavelet domain. The optimal graphs are formulated as the solutions of constrained optimization problems over sparse graphs, where the sparseness is imposed on the edges. The graphs on the wavelet domain are solved in a hierarchical layer structure. The convergence and complexity of the algorithms have been studied. Simulation experiments demonstrate that the results of our algorithms are superior to the state-of-the-art algorithms for image noise reduction and restoration.

Keywords: image restoration, image denoising, graph, ADMM, wavelet

1. Introduction

We consider the inverse problem of deriving the original image $x \in \mathcal{R}^N$ from an observation $y \in \mathcal{R}^N$, expressed as

$$y = Hx + n, \tag{1}$$

where H is an $N \times N$ matrix and $n \sim \mathcal{N}(0_{N\times 1}, \sigma I_{N\times N})$ is a vector of independent and identically distributed (i.i.d.) Gaussian random variables with standard deviation σ. We further assume that the point spread function H is known. By imposing prior information on the desired image, given as

$$\hat{x} = \arg\min_x \frac{1}{\sigma^2}\|y - Hx\|^2 + \text{penalty}(x), \tag{2}$$

where the first term is the data fidelity term for the Gaussian observation model and the second term is the regularization term, measuring the penalty of a solution that deviated away from the prior knowledge of the desired image. The modeling of the desired image is at the core of the approach [1–4]. The primary challenge of solving the problem is to recover the local high-frequency information of edges and texture in the original images that are not present in the observation.

If H is the identity matrix, problem (1) is called the noise reduction problem. The solutions vary with what type of noise is contaminated in the observation [5]. If the noise is white Gaussian noise, the state-of-the-art algorithms for the problem are BM3D [6], WBN [7], and DRUNet [8]. BM3D utilizes the tight frame representation of an image, where atoms of the frame are derived from image patches. WBN is a graphical probabilistic model of a weighted directed acyclic graph (DAG) in the wavelet domain. Different from BM3D and WBN, DRUNet is a deep learning method. It is a flexible and powerful deep CNN denoiser and the architecture is the combination of U-Net [9] and ResNet [10]. It not only outperforms the state-of-the-art deep Gaussian denoising models but also is suitable to solve plug-and-play image restoration.

If H is a blur singular matrix, problem (1) is called the image restoration problem. In the optimization-based method, the best image restoration performance both subjectively and objectively was derived from the algorithm IDD-BM3D [11]. It utilizes sparse synthetic and analytic models and de-couples the problem into blur inverse and noise reduction sub-problems, each of which is solved by a variational optimization approach. In deep learning, DPIR [8] replaces the denoising sub-problem of model-based optimization with a learning-based CNN denoiser prior which is DRUNet. By iteratively solving the data sub-problem and a prior sub-problem to restore the image.

In this chapter, we also present a restoration algorithm that combines the noise reduction algorithm with the proximal point method [12]. The primary technical contributions of our methods are the context-dependent graphical representations and the algorithms to derive the optimal graphs of each representation. Finding the optimal graph in a combinatorial way is extremely difficult and likely an NP-hard problem [13, 14]. Unlike the combinatorial approach, we impose constraints on edges and include edges in the optimal graph only when the constraints on the edges are active. This renders a computationally solvable optimization problem and the solution is a graph with only a small number of active edges.

Based on local content in an image, the context-dependent representation divides the image into singular and smooth areas. Singular areas, consisting of edges or texture, are represented and processed differently from the smooth areas. The graphs of singular areas are constructed based on the persistence and sparsity of wavelet coefficients of the image. The persistence is imposed on the inter-scale edges so that the solution at one scale can be used to confine that in adjacent scales. Meanwhile, the sparsity is imposed on the intra-scale edges that preserve the edges in which end nodes have similar intensity. In contrast, a graph of a smooth area is in the image domain and has only sparse intra-scale edges.

The algorithm to derive the optimal graphs, called graphical ADMM, is based on the alternating direction method of multipliers (ADMM) method [15, 16]. It is an efficient and robust algorithm since it breaks a complicated problem into smaller pieces, each of which is easier to handle. In our case, the node update is separated from the edge update in the optimization. In addition, for wavelet graphs, graphical ADMM approximates the multi-scale optimization problem into a sequence of sub-problems; each can be efficiently solved by convex optimization methods.

The chapter is organized as follows. In Section 2, we present the models and the construction for the context-dependent graphs. In Section 3, we formulate the noise reduction problem as a graph optimization model and present the graphical ADMM method to derive optimal graphs. In Section 4, the image restoration problem is formulated as the proximal point method that reduces the problem into a sequence of noise reduction problems, each being solved by the method in Section 3. In Section 5, experimental results and the principal differences between our and the compared methods are also discussed. Section 6 contains concluding marks.

2. Context-dependent graphical models

An image is comprised of features of edges, texture, and smooth areas. A common approach to obtain a good image processing result is to treat different features with different approaches [17, 18]. Following this approach, an image is partitioned into two types of blocks. A block containing an edge point or texture is a singular block, while the others are smooth blocks. To keep the flow, we delay the partitioning method of an image, which is described in part A of Section 5, but this is not necessary to accurately partition an image to achieve the performance demonstrated in this chapter. The singular and smooth blocks were handled with different graph optimization approaches: a singular block is in the wavelet domain, while a smooth block is in the image domain. In the wavelet domain, a singular block is represented by several weighted graphs, one corresponding to an orientation. If the wavelet transform has three orientations, LH, HL, and HH, then one graph is for LH sub-bands, another for HL sub-bands, and the third for HH sub-bands. The graph for one orientation is constructed as follows.

Each wavelet coefficient is associated with a node. Edges are comprised of inter-scale and intra-scale edges. An inter-scale edge connecting nodes in adjacent scales can direct either from a coarse scale to a finer scale or vice versa. The inter-scale edges are built-in and data-independent; they are constructed based on the wavelet persistence. In contrast, an intra-scale edge connecting nodes of the same scale is un-directed, data-dependent, and determined based on the sparseness from the graph optimization algorithm. Regularizations have been imposed on inter-scale edges to preserve the persistence of wavelet coefficients across scales and on intra-scale edges to preserve the similarity of wavelet coefficients on nodes at the two ends of an edge.

2.1 Inter-scale edges

Since wavelets can characterize singularities, representing singularities with wavelets can facilitate the restoration of edges and texture in an image. The persistence property of wavelets means that the wavelet coefficients dependency and correlations across scales. Thus, inter-scale edges were constructed to link the wavelet coefficients of the same orientation and locations at adjacent scales. Moreover, the correlations of wavelet coefficients from a coarser scale to a finer scale are different from that from a finer scale to a coarser scale. There are two types of inter-scale edges—coarse-to-fine and fine-to-coarse. The coarse-to-fine inter-scale correlation is derived based on the statistical result by Simoncelli [19], who analyzed the correlation between the dyadic wavelet coefficients in a coarse scale to those at the same location and orientation at the immediate finer scale in a natural image. The coarse-to-fine inter-scale correlation of wavelet coefficient w_{pi} at a coarse scale and wavelet coefficient w_c at the immediate finer scale can be represented in terms of minus log-probability as

$$\frac{k_1 w_i^2}{w_{pi}^2}, \tag{3}$$

where k_1 is a parameter. Thus, given the wavelet coefficients w_{pi} at the coarse scale, Eq. (3) gives the minus log probability of the wavelet coefficient w_i at the same location and orientation at the immediate fine scale.

On the other hand, the fine-to-coarse inter-scale correlation is derived from the theoretical result of wavelet singularity, analyzed by Mallat and Hwang [20]. Let w_i and w_{ci} be wavelet coefficients corresponding to the same singularity at different scales. Then, w_i and w_{ci} have the same sign and the correlation, in terms of the ratio of the modulus of wavelet coefficients, from w_{ci} at a fine scale to w_i at the immediate coarser scale can be expressed as

$$\frac{|w_i|}{|w_{ci}|} = \frac{w_i}{w_{ci}} = 2^{\alpha+\frac{1}{2}}, \tag{4}$$

where α is the Lipschitz of the singularity. If the singularity is a step edge, then α is 0. The exponent $\alpha + \frac{1}{2}$ in Eq. (4) depends on how a wavelet is normalized. Here, the wavelet is normalized to have unit 2-norm.[1] Eq. (4) can also be expressed in terms of minus log-probability as

$$k_2\left(w_i - 2^{\alpha+\frac{1}{2}} w_{ci}\right)^2, \tag{5}$$

where k_2 is a parameter. If the type α of the singularity is known, given the wavelet coefficient at the finer scale, w_{ci}, Eq. (5) gives the minus log-probability of the wavelet coefficient w_i at the coarse scale. Since step edges are the most salient features to be recovered from an image, in this chapter, we set α to 0.

2.2 Intra-scale edges

A coherent or similar structure can be used to leverage the quality of the restoration [2, 21]. This is the principle behind the success of BM3D and the example-based approach in image processing [22]. Many similarity metrics have been proposed to derive the coherent structure, such as the mutual information, the kernel functions, and the Pearson's correlation coefficient. In this chapter, the Pearson's correlation coefficient is modified for some technical concern to derive the intra-scale correlation of random variables X and Y:

$$d(X, Y) = \max\left\{0, \frac{\mathbf{E}\{(X-\mu_X)(Y-\mu_Y)\}}{\sigma_X \sigma_Y}\right\} + q, \tag{6}$$

where (μ_X, σ_X) and (μ_Y, σ_Y) are the mean and the standard deviation of X and Y, respectively, and $q > 0$ is the offset, introduced to avoid $d(X, Y) = 0$ in inequality constraints in Eq. (8). The value of $d(X, Y)$ lies in $[q, 1+q]$, measuring the similarity of

---

[1] If the wavelet is normalized to have unit 1-norm, then the exponent of Eq. (4) should be α.

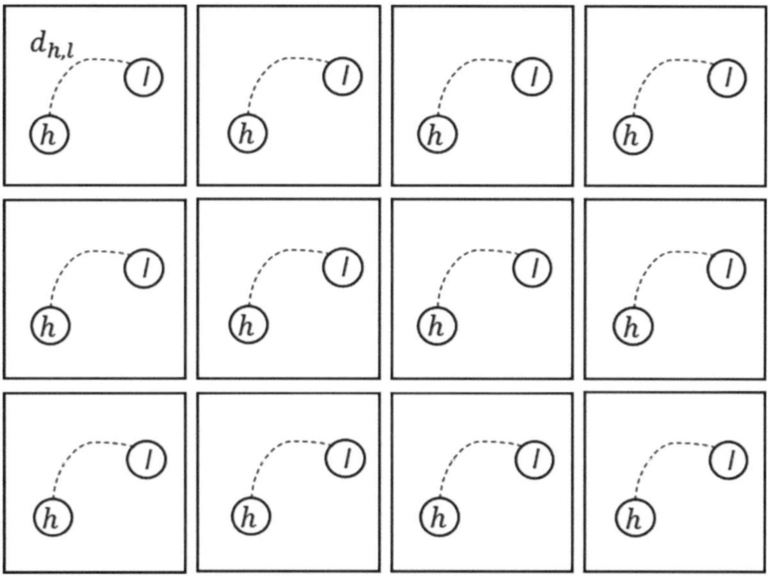

Figure 1.
An illustrative example of how the metric $d_{h,l}$ is measured. There are 12 blocks in a sub-band. All the intra-scale edges between the h-th and the l-th nodes are collected. The coefficients on the h-nodes form the random variable X and those on the l-th nodes form the random variable Y. $d_{h,l}$ is then measured based on Eq. (6).

X and Y. The smaller $d(X, Y)$ is, the more the independence of X and Y and the less likely X and Y have coherence structure. As shown in **Figure 1**, the coherence structure in an image is measured on all the coefficients of intra-scale edges in which endpoints take the same locations within a block in a sub-band.

The intra-scale edges are determined based on the sparsity constraint aiming to preserve the edges in which end nodes have similar values. The number of the edges is determined by the parameter:

$$d(y_h, y_l)|x_h - x_l| \le r, \tag{7}$$

where $d(y_h, y_l)$ is defined in Eq. (6) and obtained from the observation image, and x_h and x_l are the coefficients at the h-th and l-th nodes, respectively. If the observed values y_h and y_l are similar, the value of $d(y_h, y_l)$ is large, then $|x_h - x_l|$ would be small to satisfy the constraint. This preserves the intensities between x_h and x_h. Only the edges satisfying Eq. (7) are retained in the optimal graph. In the following, $d_{h,l}$ is used to simplify the notion $d(y_h, y_l)$ in Eq. (7).

2.3 Graph construction

The aforementioned are integrated and summarized for our context-dependent representation of an image. An image is divided into blocks. Each block is classified as either a singular block or a smooth block. A singular block is then represented with the dyadic wavelet transform, where the scale is sampled following a geometrical sequence of the ratio of 2 and the spatial domain is not down-sampled. The dyadic wavelet transform of an image is comprised of four sub-bands, LL, LH, HL, and HH, with the last three being the orientation sub-bands. A singular block is associated with

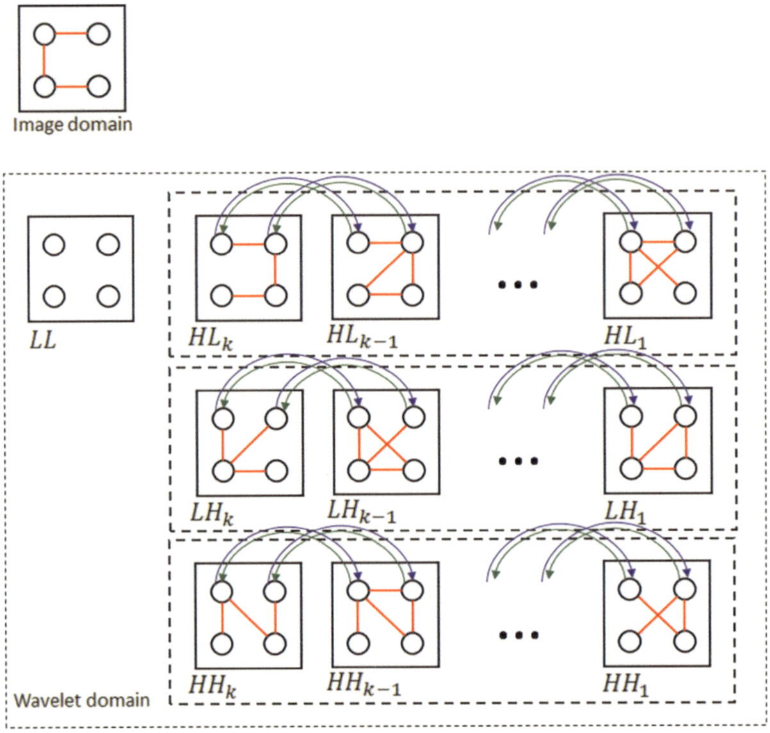

Figure 2.
A block of four pixels can be a smooth block (top) or a singular block (bottom). A smooth block is processed in the image domain. A singular block is in the wavelet domain, where a multi-scale graph is associated with each orientation. The blue and green are built-in directed inter-scale edges and the red are un-directed intra-scale edges. The inter-scale edge connects nodes at the same locations and orientation but at scales next to each other. The green edge is course-to-fine, linking a node to its parent, while the blue edge is fine to coarse, linking a node to its child. The intra-scale edges are determined by graphical ADMM, which decomposes the node update and edge update in the optimization.

three graphs—one for each orientation sub-bands. Since smooth blocks can be well restored in the image domain, the wavelet transform is not applied to the blocks and a smooth block is associated with a graph in the image domain. Each graph, associated with a singular block or a smooth block, is constructed independently of other graphs. **Figure 2** illustrates an example of graph representation for a block of four pixels.

3. Optimal graphs for noise reduction

The noise reduction problem corresponds to problem (1), where H is the identity matrix. Since each graph is solved independently of the other graphs, the following discussion is focused on one graph. A graph can be associated with a singular block or a smooth block.

3.1 Singular blocks

Let y_i and x_i be the wavelet coefficients associated to the i-th node in a sub-band of the observation image and the original image, respectively. Its parent node is denoted as $p(i)$ and child node is $c(i)$. Let $z_{h,l}$ be the variable defined on the intra-scale edge that connects the h-th and l-th nodes, which are at the same scale and orientation. The optimal wavelet graph can be derived by solving

$$\begin{cases} \min_{x_i, z_{k,l}} \left[\frac{(y_i - x_i)^2}{2\sigma^2} + \frac{k_1 x_i^2}{x_{p(i)}^2} + k_2 \left(x_i - \sqrt{2} x_{c(i)} \right)^2 \right] \\ d_{h,l} |z_{h,l}| \leq r, \\ z_{h,l} = x_h - x_l, \end{cases} \tag{8}$$

where r, σ, k_1, and k_2 are non-negative parameters and the constraints are designed as explained under Eq. (7), where r controls the number of edges in the optimal graph. If a node is at the coarsest scale 2^J, where J is the number of decomposition of the wavelet transform, it does not have a parent node and the second term in the object of Eq. (8) is zero; the third term is set to zero for a node at the finest scale 2^1 as it does not have a child node.

Problem (8) has a convenient matrix representation. Let $x = [x_i]$ and $z = [z_{h,l}]$ be the vectors of variables on nodes and intra-scale edges, respectively. Then, the linear constraints between $z_{h,l}$ and x_h and x_l in Eq. (8) can be expressed as

$$A \left[x^T z^T \right]^T = 0, \tag{9}$$

where A is a matrix with elements either -1, 1, or 0. Let $A = [A_x A_z]$. If follows that

$$A \left[x^T z^T \right]^T = A_x x + A_z z. \tag{10}$$

Each row of A_x has one element which value is 1 and another element which value is -1 and the rest of value 0. Meanwhile, each row of has one element of values and the rest of value . Let $\lambda = [\lambda_{h,l}] \geq 0^2$ and μ be the vectors of Lagrangian variables associated to inequality constraints and the equality constraints in Eq. (8), respectively. The augmented Lagrangian of Eq. (8) is

$$\begin{aligned} L_\rho(x, z, \lambda, \mu) = & \sum_i \left[\frac{(y_i - x_i)^2}{2\sigma^2} + \frac{k_1 x_i^2}{x_{p(i)}^2} + k_2 \left(x_i - \sqrt{2} x_{c(i)} \right)^2 \right] \\ & + \mu^T (A_x x + A_z z) + \frac{\rho}{2} \| A_x x + A_z z \|^2 + \sum_{h,l} \lambda_{h,l} d_{h,l} |z_{h,l}| - r \mathbf{1}^T \lambda, \end{aligned} \tag{11}$$

where **1** is a vector with all members being 1; and $\rho > 0$ are fixed parameters.

The ADMM algorithm intends to blend the decomposability of dual ascent with the superior convergence properties of the method of multipliers. Here, ADMM is used to derive the optimal graph by separating the node and edge update. Graphical ADMM derives the saddle points of

$$\max_{\lambda \geq 0, \mu} \min_{x, z} L_\rho(x, z, \lambda, \mu) \tag{12}$$

---

[2] Let $\lambda = [\lambda_{h,l}]$. Then, $\lambda \geq 0$ if and only if $\lambda_{h,l} \geq 0$ for all h and l.

by iterating the following updates:

$$\begin{cases} x^{k+1} = \arg\min_{x} L_\rho\left(x, z^k, \lambda^k, \mu^k\right); \\ z^{k+1} = \mathbb{Q}\left(\arg\min_{z} L_\rho\left(x^{k+1}, z, \lambda^k, \mu^k\right)\right); \\ \lambda_{h,l}^{k+1} = \mathbb{P}_{\geq 0}\left(\lambda_{h,l}^k - \varepsilon\left[r - d_{h,l}|z_{h,l}^{k+1}|\right]\right); \\ \mu^{k+1} = \mu^k + \rho\left(A_x x^{k+1} + A_z z^{k+1}\right); \end{cases} \tag{13}$$

where $\mathbb{Q}$ and $\mathbb{P}_{\geq 0}$ are the orthogonal projections to satisfy constraints $d_{j,l}\|z_{h,l}\|_1 \leq r$ and $\lambda_{h,l} \geq 0$, respectively; and $\varepsilon > 0$ is the stepping size for the dual ascent of Lagrangian variables λ. The first and second updates in Eq. (13) update the node variables and edge variables, respectively. The update of the dual variables λ is derived based on the necessary conditions at an optimum of Eq. (8) that

$$\begin{cases} \lambda_{h,l} \geq 0; \\ \lambda_{h,l}\left(r - d_{h,l}|z_{h,l}^{k+1}|\right) = 0. \end{cases} \tag{14}$$

The third update in Eq. (13) has the following interpretation. If $\left(r - d_{h,l}|z_{h,l}^{k+1}|\right) > 0$, then $\lambda_{h,l}^k$ will decrease and keep its value to be non-negative by $\mathbb{P}$. The value of $\lambda_{h,l}$ can be repeatedly decreased by increasing the iteration number k until either $\lambda_{h,l} = 0$ or $\left(r - d_{h,l}|z_{h,l}^{k+1}|\right) = 0$, where the edge is active, and the optimal conditions (14) satisfy. At optimum, either the Lagrangian associated with an edge is zero or the constraint on the edge is active. Only the active edges are retained in the graph. Since the number of active edges is sparse, the edges in the optimal graph are sparse. The solutions for the updates rules for primal variables are derived in Sections 3.1 and 3.2, respectively.

3.2 Smooth blocks

A smooth block is processed in the image domain, where each pixel is associated with a node in a graph. Problem (8) becomes finding the optimal graph by solving

$$\begin{cases} \min_{x_i, z_{h,l}} \sum_i \frac{(y_i - x_i)^2}{2\sigma^2} \\ d_{h,l}|z_{h,l}| \leq r, \\ z_{h,l} = x_h - x_l. \end{cases} \tag{15}$$

The optimal graph can be derived by a method similar to that for a singular block.

3.3 Update edges

The update rule for edge variables z is to solve, with fixed $x, \lambda,$ and μ,

$$\begin{cases} \min_z L_\rho(x, z, \lambda, \mu) \\ d_{h,l}|z_{h,l}| \leq r. \end{cases} \tag{16}$$

If only the terms in Eq. (3) relevant to the optimization variables, $\{z_{h,l}\}$, are concerned, Eq. (16) becomes

$$\begin{cases} \min \sum_{h,l} \left(-\mu_{h,l}^T z_{h,l} + \frac{\rho}{2} (x_h - x_l - z_{h,l})^2 + \lambda_{h,l} d_{h,l} |z_{h,l}| \right) \\ |z_{h,l}| \leq \frac{r}{d_{h,l}}, \end{cases} \tag{17}$$

because $d_{h,l}$ is non-zero as defined in Eq. (6). Equation (17) indicates that each edge variable can be updated independently of the others.

We first solve Eq. (17) without the constraint and obtain the solution $u_{h,l}$ of

$$\arg\min_{z_{h,l}} \left(-\mu_{h,l}^T z_{h,l} + \frac{\rho}{2} (x_h - x_l - z_{h,l})^2 + \lambda_{h,l} d_{h,l} |z_{h,l}| \right), \tag{18}$$

by soft-thresholding with

$$u_{h,l} = \begin{cases} x_h - x_l + \frac{\mu_{h,l}}{\rho} - \frac{\lambda_{h,l}}{\rho} d_{h,l}, & \text{if } x_h - x_l + \frac{\mu_{h,l}}{\rho} \geq \frac{\lambda_{h,l}}{\rho} d_{h,l}; \\ x_h - x_l + \frac{\mu_{h,l}}{\rho} + \frac{\lambda_{h,l}}{\rho} d_{h,l}, & \text{if } x_h - x_l + \frac{\mu_{h,l}}{\rho} \leq -\frac{\lambda_{h,l}}{\rho} d_{h,l}; \\ 0, & \text{otherwise.} \end{cases} \tag{19}$$

It is then followed by orthogonally projecting $u_{h,l}$ to satisfy the constraint by solving

$$\begin{cases} \min_{z_{h,l}} \frac{1}{2} (z_{h,l} - u_{h,l})^2 \\ |z_{h,l}| \leq \frac{r}{d_{h,l}}. \end{cases} \tag{20}$$

Eq. (20) can be solved by a sequence of soft-thresholding operations. The algorithm is sketched as follows. First, we check whether $|u_{h,l}| \leq \frac{r}{d_{h,l}}$. If it is, $u_{h,l}$ is the solution. Otherwise, we begin with a small γ and solve

$$z_{h,l}^+ = \arg\min_{z_{h,l}} \frac{1}{2} (z_{h,l} - u_{h,l})^2 + \gamma |z_{h,l}|. \tag{21}$$

The solution is the soft-thresholding as

$$z_{h,l}^+ = \begin{cases} 0, & \text{if } |u_{h,l}| \leq \gamma; \\ \left(1 - \frac{\gamma}{|u_{h,l}(i)|}\right) u_{h,l}, & \text{otherwise.} \end{cases} \tag{22}$$

If $|z_{h,l}^+| \leq \frac{r}{d_{h,l}}$, $z_{h,l}$ is updated to $z_{h,l}^+$, the algorithm stops. Otherwise, γ is increased and Eq. (22) is solved again. Since increasing γ decreases $|z_{h,l}^+|$, this algorithm always stops and updates the edge variable $z_{h,l}$ to meet the constraint.

The complexity to update edge variables is analyzed. If the number of pixels of a block is n and if the dyadic wavelet transform takes J scales, then the number of edge constraints on a singular block is $O(Jn^2)$ and the number of edge constraints on a smooth block is $O(n^2)$. Let K_1 be the maximum number of iterations to derive the solution for Eq. (20) for all graphs. The complexity of one edge update is $O(K_1 n^2(|J_F|+3|J_W|J))$, where 3 is the number of orientations, $|J_F|$ and $|J_W|$ are numbers of singular blocks and smooth blocks, respectively.

3.4 Update nodes

The node update for a singular block is more complicated than that for a smooth block because a graph for a singular block has a multi-scale structure, where adjacent scales are linked by inter-scale edges.

3.5 Singular blocks

To update the nodes x in a singular block is to solve the augmented Lagrangian function (3) via

$$\arg\min_x L_\rho(x, z, \lambda, \mu), \tag{23}$$

for given z, λ, and μ. This is not a convex problem because the second term in Eq. (3) is non-convex.

Our approach is to decompose the problem based on the scale parameter into a sequence of sub-problems. Each scale is associated with two convex sub-problems: one is a coarse-to-fine sub-problem and the other is a fine-to-coarse sub-problem. The coarse-to-fine sub-problem assumes the parent nodes at scale 2^{s+1} were updated earlier, while the fine-to-coarse sub-problem assumes the child nodes at scale 2^{s+1} were updated earlier. Let k be the current iteration number. The course-to-fine sub-problem updates the nodes at scale 2^s by minimizing[3]

$$\sum_i \left[\frac{(y_i - x_i)^2}{2\sigma^2} + \frac{k_1 x_i^2}{\left[x_{p(i)}^k\right]^2} \right] + \sum_{h,l} \left[\mu_{h,l}^{k-1} \left(z_{h,l}^{k-1} - (x_h - x_l)\right) + \frac{\rho}{2} \left(z_{h,l}^{k-1} - (x_h - x_l)\right)^2 \right]. \tag{24}$$

On the other hand, the fine-to-coarse sub-problem updates the nodes at scale by minimizing[4]

$$\sum_i \left[\frac{(y_i - x_i)^2}{2\sigma^2} + k_2 \left(x_i - \sqrt{2} x_{c(i)}^k\right)^2 \right] + \sum_{h,l} \left[\mu_{h,l}^{k-1} \left(z_{h,l}^{k-1} - (x_h - x_l)\right) + \frac{\rho}{2} \left(z_{h,l}^{k-1} - (x_h - x_l)\right)^2 \right]. \tag{25}$$

The node update problem (23) can then be approximated by repeatedly solving the coarse-to-fine iteration followed by the fine-to-coarse iteration. The coarse-to-fine iteration solves a sequence of the coarse-to-fine sub-problems beginning at the

---

[3] The second term at below is zero, when 2^s is the coarsest scale.

[4] The second term below is zero when nodes are at the finest scale.

coarsest scale. In contrast, the fine-to-coarse iteration solves a sequence of the fine-to-coarse sub-problems beginning at the finest scale.

Problems (24) and (25) can be efficiently solved. The objectives in the sub-problems are strictly convex functions because their Hessian matrices are positive definite (as can be observed from the inverse matrix of Eqs. (26) and (27)) and, thus, the optimal solution of each is unique. The closed-form solutions of the sub-problems can be derived as follows.

For convenience, we omit all the superscript index in Eqs. (24) and (25) and let A_x^s and A_z^s denote the sub-matrices of A_x and A_x, respectively. A_x^s and A_z^s retain only the rows and columns in A_x and A_x corresponding to the nodes and edges at scale 2^s, respectively. We also let x_s, z_s, and μ_s denote the vectors of nodes, edges, and Lagrangian variables at scale 2^s. The closed-form solution of x_s of the coarse-to-fine sub-problem is

$$\begin{cases} \left[\frac{1}{\sigma^2}I + C_{s+1}^T C_{s+1} + \rho\left(A_x^s\right)^T A_x^s\right]^{-1}\left(\frac{1}{\sigma^2}y_s - \left(A_x^s\right)^T\left(\mu_s - \rho A_z^s z_s\right)\right) \text{for } s \neq J; \\ \left[\frac{1}{\sigma^2}I + \rho\left(A_x^s\right)^T A_x^s\right]^{-1}\left(\frac{1}{\sigma^2}y_s - \left(A_x^s\right)^T\left(\mu_s - \rho A_z^s z_s\right)\right) \text{for } s = J; \end{cases} \tag{26}$$

where C_{s+1} is a diagonal matrix which diagonal element at (i, i) is $\frac{\sqrt{k_1}}{\sqrt{2}\|x_{p(i)}\|}$ and 2^J is the coarsest scale. On the other hand, the closed-form solution of x_s for the fine-to-coarse sub-problem is

$$\begin{cases} \left[\left(\frac{1}{\sigma^2} + 2k_2\right)I + \rho\left(A_x^s\right)^T A_x^s\right]^{-1}\left(\frac{1}{\sigma^2}y_s + 2\sqrt{2}k_2 x_{s-1} - \left(A_x^s\right)^T\left(\mu_s - \rho A_z^s z_s\right)\right) \text{for } s \geq 2; \\ \left[\frac{1}{\sigma^2}I + \rho\left(A_x^s\right)^T A_x^s\right]^{-1}\left(\frac{1}{\sigma^2}y_s - \left(A_x^s\right)^T\left(\mu_s - \rho A_z^s z_s\right)\right) \text{for } s = 1; \end{cases} \tag{27}$$

The complexity of the matrix inversion in Eqs. (26) and (27) is low since A_x^s is a sparse matrix and each row of A_x^s has at most one 1 and one -1 and the rest are zero. The complexity of the sparse matrix inversion in Matlab is proportional to the number of non-zero elements in the matrix. Thus, one iteration of either coarse-to-fine or fine-to-coarse of a graph takes the complexity $O(Jn)$, where J is the number of decomposition and n is the number of pixels at a scale.

3.6 Smooth blocks

The node update for a smooth block can be analytically derived from the problem (15) at the condition that $\{z_{h,l}\}$ is given. If the superscript index is omitted, the closed-form solution is

$$\left[\frac{1}{\sigma^2}I + \rho A_x^T A_x\right]^{-1}\left(\frac{1}{\sigma^2}y - \left(A_x^s\right)^T(\mu - \rho A_z z)\right), \tag{28}$$

where A_x and A_z are defined in Eq. (10). The complexity of the inversion of the sparse matrix $\frac{1}{\sigma^2}I + \rho A_x^T A_x$ is propositional to the non-zero elements in the matrix, which is $O(n)$, where n is number of pixels in a block.

If an image has $|J_W|$ singular blocks and $|J_F|$ smooth blocks, and if J is the number of wavelet decompositions, the total complexity of node updates of the image is $O(n(3K_2J|J_W| + |J_F|))$, where K_2 is the maximum number of iterations of coarse-to-fine and fine-to-coarse node update for a singular block.

Graphical ADMM consists of a sequence of updating the primal and dual variables and the complexity of the algorithm is dominated by the primal variable updates. Our analysis of one iteration of Eq. (13) for node updates and edge updates indicates that the costs are $O(n(3K_2J|J_W| + |J_F|))$ and $O(K_1n^2(3|J_W|J+|J_F|))$, respectively, where n is the number of pixels in a block.

4. Optimal graphs for image restoration

There are various image restoration methods [23]. Here, we use the proximal approach proposed in [24] and [12]. The method smartly reduces the image restoration problem into a sequence of noise reduction problems. Since graphical ADMM for noise reduction is efficient, it can be adopted to derive the optimal graphs for image restoration. Like for noise reduction, a graph is handled independently of the other graphs. The following discussion is focused on deriving the optimal graph for a block.

Let $h(x)$ be the objective function

$$\frac{1}{2}\|y - Hx\|^2, \tag{29}$$

with a known blur kernel H; x_0 is the vector of the current restored image. The proximity function is defined as

$$d_H(x, x_0) = \frac{\beta}{2}\|x - x_0\|^2 - \frac{1}{2}\|Hx - Hx_0\|^2. \tag{30}$$

The parameter β is chosen so that $d_H(x, x_0)$ is strictly convex with respect to x. This implies that its Hessian $\beta I - H^T H$ is a positive definite matrix, which can be achieved by choosing $\beta > \lambda_{\max}(H^T H)$ (the maximal eigenvalue of the matrix $H^T H$). The proximal objective is defined as

$$\tilde{h}(x, x_0) = h(x) + d_H(x, x_0). \tag{31}$$

Simplifying the above objective, we have the following simpler form by removing $\|Hx\|$ from the proximal objective as

$$\tilde{h}(x, x_0) = \frac{\beta}{2}\|x - \left(x_0 + \frac{1}{q}H^T(y - Hx_0)\right)\|^2 + K \tag{32}$$

where K contains terms unrelated to x. Since x_0, q, H, y are given, the proximal objective can be regarded as a noise reduction problem with the observation vector, $x_0 + \frac{1}{q}H^T(y - Hx_0)$. Thus, $\tilde{h}(x, x_0)$ can be the first term in noise reduction problem (8) and the algorithm for noise reduction can be used to derive the optimal graph via separating the node and edge updates following Eq. (13) and procedures in Section 3.

5. Experiments and comparisons

We consider several image denoising and deblurring scenarios used as the benchmarks in state-of-the-art algorithms for performance evaluations and comparisons. The setting of experiments is given as follows. The experiments were conducted on images in Sets I and II in **Figure 3**. Set I contains six gray-scaled natural images, Einstein, Boat, Barbara, Lena, Cameraman, and House. The size of each image is 512×512 or 256×256, downloaded from the USC-SIPI image database [25]; and Set II contains six gray-scaled textures. Some of them were taken from the Brodatz texture set. Through all experiments, each image is divided into 16 equal-sized blocks. A singular block is decomposed into four scales dyadic wavelet transform with the CDF 9/7 wavelet filters. Since the CDF 9/7 filters are close to orthogonal wavelet filters, the noise variance at any sub-band can be set to σ^2, the variance of noise in the image domain [7].

5.1 Noise reduction performance

Our noise reduction performance was compared against that of BM3D, WBN, and DRUNet. The perceptual quality of the methods is shown in **Figure 4**. The Lena image of BM3D over-smooths the highlighted area of hat, which is rich in edges and textures. Similarly, textures in the highlighted area of hat in DRUNet are smooth. The image of WBN, on the contrary, under-smooths the highlighted smooth area around the chin and shoulder of Lena. These artifacts have been amended by graphical ADMM, as shown in **Figure 4f**.

The quantity comparisons, measured by the peak-signal-to-noise ratio (PSNR), of Set I and Set II are shown in **Tables 1** and **2**, respectively. The testing environments were images contaminated with the noise of variances, σ^2. As shown, the deep learning-based method (DRUNet) achieves the highest score in almost all environments. However, in the optimization-based methods (BM3D, WBN, proposed), graphical ADMM achieves unanimously the highest score in all environments.

Figure 3.
The images in set I (first row) and set II (second row). Set I contains six gray-scaled natural images and set II contains six gray-scaled textures.

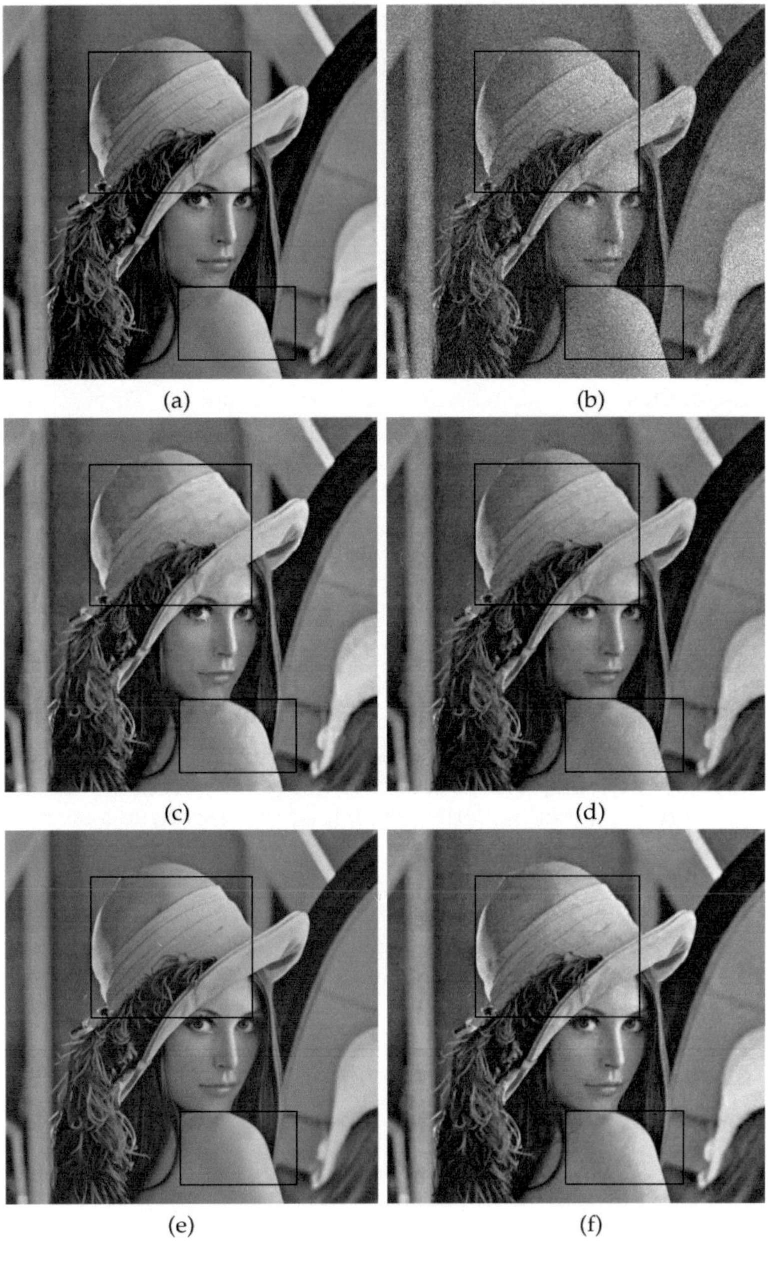

Figure 4.
Comparisons of the denoised Lena images derived by BM3D, WBN, DRUNet, and graphical ADMM. The noise standard deviation is set at $\sigma = 25$: (a) the original 512×512 Lena image; (b) the noised image; (c) the result of BM3D; (d) the result of WBN; (e) the result of DRUNet; and (f) the result of graphical ADMM. Graphical ADMM preserves both the smooth and edged areas in the original image, as shown in the highlighted areas.

5.2 Image restoration performance

Table 3 presents five-point spread functions (PSFs) used for image restoration in literature [11]. Each PSF was normalized to have unit 1-norm before it was used to

Image	Method	PSNR					
		$\sigma = 10$	$\sigma = 15$	$\sigma = 20$	$\sigma = 25$	$\sigma = 30$	$\sigma = 35$
Einstein	BM3D	34.4392	33.0331	32.1694	31.4186	30.8709	30.3777
	WBN	34.4848	33.0821	32.3429	31.4728	30.9178	30.4109
512 × 512	Proposed	34.5013	33.1005	32.3544	31.4862	30.9306	30.4255
	DRUNet	34.9948	33.6019	32.7411	32.1092	31.5952	31.1622
	BM3D	33.8883	32.1067	30.8554	29.8356	29.0954	28.2992
Boat	WBN	33.9095	32.1369	30.8874	29.8561	29.1273	28.3285
512 × 512	Proposed	33.9241	32.1504	30.9063	29.8772	29.1416	28.3410
	DRUNet	34.4264	32.7123	31.5194	30.5768	29.8391	29.1826
	BM3D	34.9567	33.0666	31.7376	30.7176	29.7049	28.8879
Barbara	WBN	34.9643	33.0831	31.7515	30.7332	29.7233	28.4571
512 × 512	Proposed	34.9704	33.1012	31.7735	30.7468	29.7458	28.4675
	DRUNet	35.2115	33.4389	32.1951	31.2341	30.4275	29.7520
	BM3D	36.6367	34.8782	33.0567	32.5501	31.6531	31.0301
Lena	WBN	36.6354	34.8886	33.3048	32.4488	31.5617	30.9148
512 × 512	Proposed	36.6447	34.8960	33.3065	32.5632	31.6744	31.1022
	DRUNet	36.4431	34.9269	33.8363	32.9669	32.2285	31.6072
	BM3D	34.1355	31.8449	30.3797	29.4118	28.5516	27.8758
Cameraman	WBN	34.1637	31.8675	30.5629	29.5723	28.7274	28.0487
256 × 256	Proposed	34.1668	31.8776	30.5732	29.5804	28.7335	28.0614
	DRUNet	34.9927	32.9133	31.5788	30.6079	29.8462	29.2131
	BM3D	36.6638	34.9028	33.7349	32.9084	32.1240	31.5103
House	WBN	36.8538	34.9302	33.7620	32.9363	32.1571	31.5390
256 × 256	Proposed	36.8665	34.9367	33.7742	32.9420	32.1643	31.5404
	DRUNet	37.4420	35.8267	34.7084	33.9251	33.2517	32.6733

Table 1.
Comparisons of the PSNRs of the noise reduction methods on noisy images with the noise of standard deviation σ in set I.

Image	Method	PSNR					
		$\sigma = 10$	$\sigma = 15$	$\sigma = 20$	$\sigma = 25$	$\sigma = 30$	$\sigma = 35$
F0	BM3D	32.5200	30.3213	28.8595	27.7279	26.7751	25.9243
	WBN	32.6212	30.4143	28.9314	27.8087	26.8644	26.0136
512 × 512	Proposed	32.6338	30.4365	28.9572	27.8276	26.8867	26.0345
	DRUNet	33.7186	31.7164	30.2770	29.1838	28.3420	27.6675
	BM3D	29.3575	26.3832	24.3540	22.8154	21.6195	20.6102
F3	WBN	29.5361	26.5738	24.5363	23.0236	21.8514	20.9144

Image	Method	PSNR					
		$\sigma = 10$	$\sigma = 15$	$\sigma = 20$	$\sigma = 25$	$\sigma = 30$	$\sigma = 35$
512 × 512	Proposed	29.5464	26.5896	24.5554	23.0467	21.8722	20.9364
	DRUNet	30.0832	27.2371	25.3141	23.8812	22.7516	21.8235
	BM3D	29.8987	27.3043	25.6216	24.3522	23.4010	22.5878
F7	WBN	30.1253	27.3765	25.6782	24.4186	23.4862	22.6574
512 × 512	Proposed	30.1470	27.3932	25.6914	24.4375	23.5002	22.6786
	DRUNet	30.8467	28.2704	26.5578	25.3201	24.3751	23.6274
	BM3D	34.0075	32.4024	31.2442	30.311	29.5213	28.7919
g3	WBN	34.0384	32.4566	31.2517	30.3472	29.5506	28.8064
512 × 512	Proposed	34.0415	32.4734	31.2570	30.3613	29.5712	28.8115
	DRUNet	34.2632	32.7016	31.6367	30.7665	30.0146	29.3556
	BM3D	31.3155	29.0087	27.4678	26.3278	25.4362	24.7034
p3	WBN	31.3385	29.0274	27.4936	26.3414	25.4535	24.7153
512 × 512	Proposed	31.3520	29.0475	27.5102	26.3588	25.4763	24.7274
	DRUNet	31.9992	29.7159	28.1847	27.0464	26.1412	25.3947
	BM3D	31.5652	29.5789	28.3481	27.4696	26.7822	26.2039
r3	WBN	31.6362	29.6675	28.4277	27.5486	26.8774	26.2745
512 × 512	Proposed	31.6521	29.6836	28.4672	27.5665	26.9013	26.3020
	DRUNet	31.8651	29.9371	28.7082	27.8162	27.1131	26.5323

Table 2.
Comparisons of the PSNRs of the noise reduction methods on set II texture images.

Blur Kernel	Formulation	size
h_1	$\frac{1}{1+d_x^2+d_y^2}$	15 × 15
h_2	Uniform	9 × 9
h_3	$\begin{bmatrix}1\\4\\6\\4\\1\end{bmatrix} \times [1,4,6,4,1]$	5 × 5
h_4	Gaussian ($\sigma = 1.6$)	25 × 25
h_5	Gaussian ($\sigma = 0.4$)	25 × 25

Table 3.
Blur kernels for experiments. The d_x and d_y in h_1 are, respectively, the horizontal and vertical distances of a pixel to the center of the blur kernel.

blur an image. The performance was compared with the state-of-the-art methods, IDD-BM3D and DPIR. To have fair comparisons, both methods used the same initial images in each experiment. The visual quality of the restored images is shown in **Figure 5** and the blue and red boxes are magnifications of the highlighted areas in the

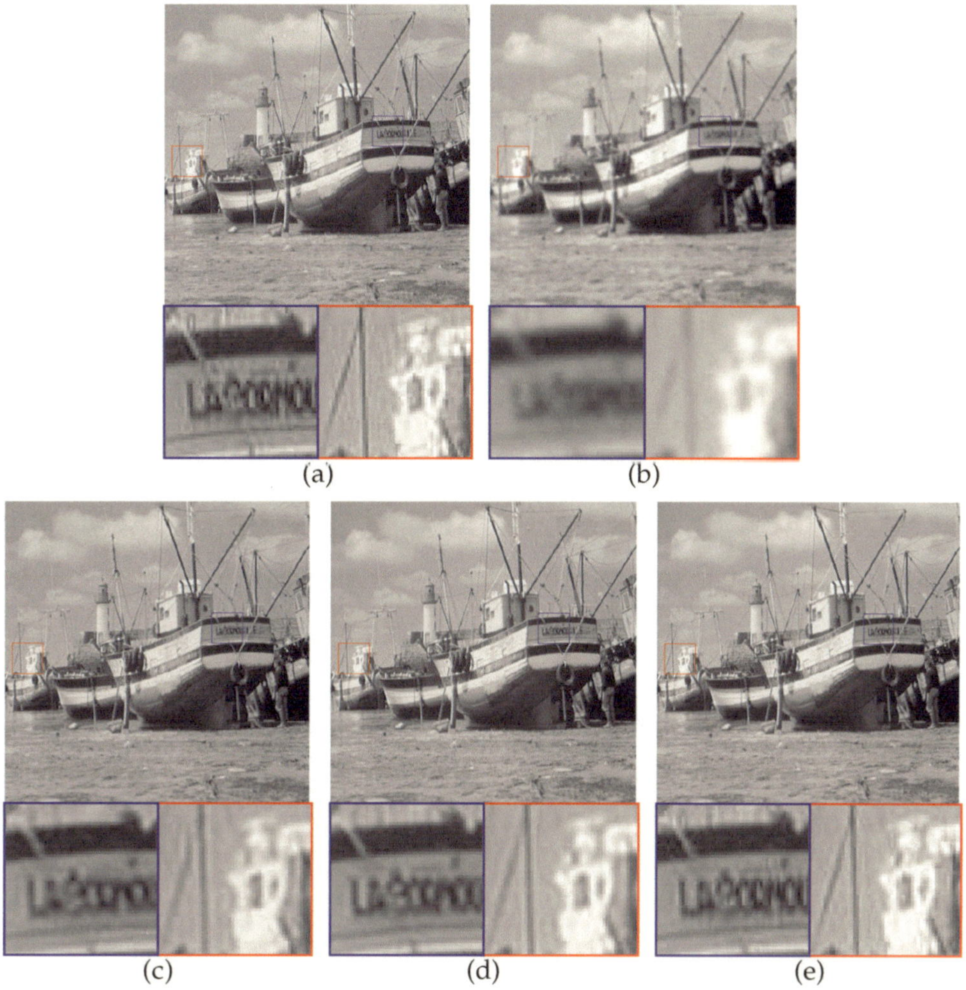

Figure 5.
Comparisons of the deblurred images. The blue and red boxes are the magnified areas in the image. (a) the original 512×512 boat image; (b) the blurred image with blur kernel h_4; (c) the image of IDD-BM3D; (d) the image of DPIR; and (e) the image of graphical ADMM. The overall perceptual quality of our image is better since that of IDD-BM3D and DPIR are over-smoothed.

image. Compared with the original images, the overall perceptual quality of the images of IDD-BM3D and DPIR appear over-smoothed, whereas graphical ADMM preserves more image details, leading to better perceptual quality. Graphical ADMM can preserve more details because it uses the multi-scale approach in treating the texture and edge regions. The wavelet persistence property allows information at coarse scales to pass to fine scales and vice versa. As a result, graphical ADMM yields shaper results in recovering singular points in images.

The quantity comparison is shown in **Figure 6**, where the performance improvement of graphical ADMM over IDD-BM3D was measured by the ISNR (increased signal-to-noise ratio) [26]. The ISNR quantitatively assesses the restored images with known ground truths. Let y, x, and x_0 be the vector representations of the observation, the restored image, and the ground truth, respectively; the ISNR is defined as

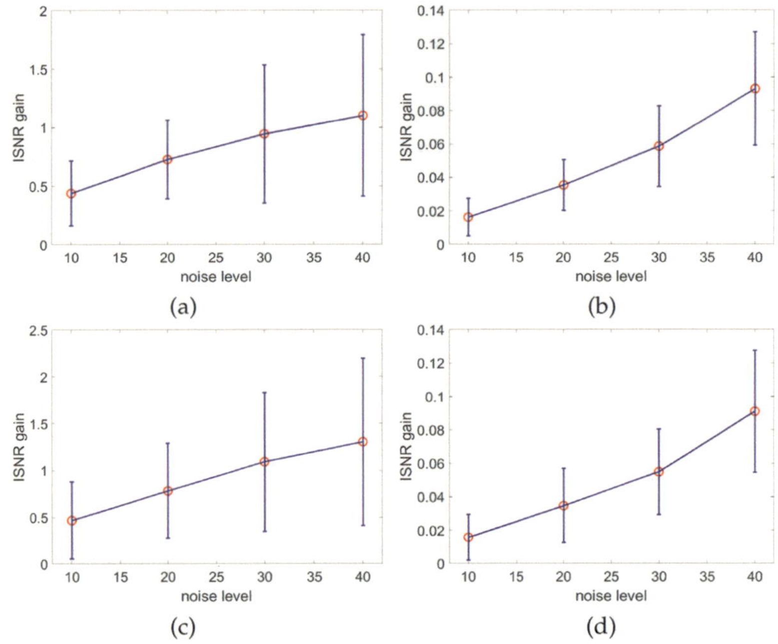

Figure 6.
Average and standard deviation of the ISNR gain of DPIR over IDD-BM3D and graphical ADMM over IDD-BM3D. Each image is blurred and then added to white noise of standard deviation indicated by the noise level. The image was then deblurred. An ISNR gain was calculated from the de-blurred images. The circled point and bar of a measurement at a noise level are the average and standard deviation, respectively, of thirty ISNR gains of natural images from set I ((a) DPIR over IDD-BM3D and (b) graphical ADMM over IDD-BM3D) and texture images from set II ((c) DPIR over IDD-BM3D and (d) graphical ADMM over IDD-BM3D). As shown, the curves of ISNR gain increase steadily and progressively when the noise level increases.

$$10 \log_{10} \frac{\|y - x_0\|^2}{\|x - x_0\|^2}. \tag{33}$$

The higher the ISNR value of a restored image, the better the restoration quality of the image. The ISNR gain of graphical ADMM over that of IDD-BM3D is defined as

$$\text{ISNR(graphical ADMM)-ISNR(IDD-BM3D)}, \tag{34}$$

and the ISNR gain of DPIR over that of IDD-BM3D is defined as

$$\text{ISNR(DPIR)-ISNR(IDD-BM3D)}. \tag{35}$$

Figure 6a and **c** show the ISNR gain of DPIR over IDD-BM3D in Set I and Set II, respectively. **Figure 6b** and **d** show the ISNR gain of graphical ADMM over IDD-BM3D in Set I and Set II, respectively. Let us take **Figure 6b** as an example. At a noise level, each image in Set I was first blurred by a kernel in **Table 3**. The result was added to white noise to obtain a noisy blurred image. This procedure generated thirty noisy blurred images since Set I contains six images and **Table 3** has five blur kernels. Each noisy blurred image was deblurred. The ISNR gain of the image obtained by graphical ADMM and that by IDD-BM3D was calculated. The thirty ISNR gains were then used to calculate the mean and standard derivation, as shown in **Figure 6**. The mean ISNR gain of graphical ADMM increases steadily and progressively over IDD-BM3D, as the noise level increases.

5.3 Discussions

DRUNet and DPIR are deep learning methods and the training data with a noise level of σ ranges from 0 to 50. In the experiments, they have the best performance in quantity comparisons but graphical ADMM is the best in visual quality. For learning-based methods, the training data is important and related to the performance. The inferred results are data-driven and not interpretable. If the training data is less or the distribution of the testing data is not similar to training, the performance will be worse. If the artifacts occurred in the results, we do not know how it happened because the network is just composed of many coefficients trained from the training data. At the same time, the time cost for training is very high. These are all the drawbacks of learning-based methods. However, the optimization-based methods are not limited to the training data. The results are derived from the objective function and interpretable. In addition, they are more stable in practical applications. So, there is a trade-off between learning and optimization-based methods.

For the optimization-based methods, the experiments have demonstrated the advantages of graphical ADMM in both the noise reduction and image restoration tasks over the compared methods. Recall that BM3D and IDD-BM3D adopt the image-dependent tight frame representations. IDD-BM3D also combines the analytic and synthetic optimization methods by de-coupling the noise reduction problem and the image restoration problem. This yields a game-theoretical approach that two formulations are used to minimize a single objective function. The solution adopted by IDD-BM3D is a Nash equilibrium point. The WBN represents an image as a multi-scale probabilistic DAG and adopts the belief propagation to derive the MAP solution.

The advantages of graphical ADMM lie in the context-dependent decompositions of an image horizontally in space and vertically along the scales in handling the image details. The spatial decomposition allows our method to overcome the cons of under-smoothing the smooth areas in WBN and keeps the pros of WBN that preserves sharp edges. Meanwhile, graphical ADMM is much more efficient than the time-consuming belief propagation adopted in WBN.

The mixture of data-dependent and data-independent edges in wavelet graph construction is a significant feature of our method. The intra-scale edges are determined by a data-dependent adaptive process, which imposes sparseness by keeping the edges which end nodes have similar coefficients in the optimal graph. The inter-scale edges are data-independent, built-in to leverage the wavelet persistence property. The inter-scale edges, passing information of singularities from finer scales to coarser scales and vice versa, can preserve more texture and edges in original images. This distinguishes our algorithm from BM3D and IDD-BM3D, which encode structure in atoms of a dictionary and select a few atoms for image representation.

6. Conclusions

We present a novel approach by combining spatial decomposition, vertical (multi-scale) decomposition, and ADMM optimization in a graphical framework for image noise reduction and restoration tasks. The graphical ADMM method has demonstrated that its results are superior to those of state-of-the-art algorithms. We also demonstrated that mixing data-dependent and data-independent structures in a graph representation can leverage the sparseness and persistence of a wavelet representation. Rather than adopting a combinatorial approach to derive an optimal graph, we showed that the graph can be derived by a numerically tractable optimization approach. In addition, we showed that the optimization problem is well coupled with our graph representation, and can be decomposed into a sequence of convex subproblems, with each having an efficient closed-form solution. This opens a new perspective of combining a mixture of data-adaptive and data-independent structures, hierarchical decomposition, and optimization algorithms in modeling, representing, and solving more image processing tasks.

Author details

Jinn Ho, Shih-Shuo Tung and Wen-Liang Hwang*
Institute of Information Science, Academia Sinica, Taiwan

*Address all correspondence to: whwang@iis.sinica.edu.tw

References

[1] Rudin L, Osher S, Fatemi E. Nonlinear total variation based noise removal algorithms. Physica D: Nonlinear Phenomena. 1992;**60**(1–4):259-268. DOI: 10.1016/0167-2789(92)90242-F

[2] Buades A, Coll B, Morel J. A review of image denoising algorithms, with a new one. Multiscale Modeling & Simulation. 2005;**4**(2):490-530. DOI: 10.1137/040616024

[3] Elad M, Aharon M. Image denoising via sparse and redundant representations over learned dictionaries. IEEE Transactions on Image Processing. 2006;**15**(12):3736-3745. DOI: 10.1109/TIP.2006.881969

[4] Chatterjee P, Milanfar P. Is denoising dead? IEEE Transactions on Image Processing. 2010;**19**(4):895-911. DOI: 10.1109/TIP.2009.2037087

[5] Li J, Shen Z, Yin R, Zhang X. A reweighted l2 method for image restoration with poisson and mixed poisson-gaussian noise. Inverse Problem and Imaging. 2015;**9**(3):875-894. DOI: 10.3934/ipi.2015.9.875

[6] Dabov K, Foi A, Katkovnik V, Egiazarian K. Image denoising by sparse 3-D transform-domain collaborative filtering. IEEE Transactions on Image Processing. 2007;**16**(8):2080-2095. DOI: 10.1109/TIP.2007.901238

[7] Ho J, Hwang W. Wavelet Bayesian network image denoising. IEEE Transactions on Image Processing. 2013;**22**(4):1277-1290. DOI: 10.1109/TIP.2012.2220150

[8] Zhang K, Li Y, Zuo W, Zhang L, Gool L, Timofte R. Plug-and-play image restoration with deep denoiser prior. IEEE Transactions on Pattern Analysis and Machine Intelligence. 2021. DOI: 10.1109/TPAMI.2021.3088914

[9] Ronneberger O, Fischer P, Brox T. U-net: Convolutional networks for biomedical image segmentation. International Conference on Medical Image Computing and Computer-Assisted Intervention. 2015:234-241. DOI: 10.1007/978-3-319-24574-4_28

[10] He K, Zhang X, Ren S, Sun J. Deep residual learning for image recognition. IEEE Conference on Computer Vision and Pattern Recognition. 2016:770-778. DOI: 10.1109/CVPR.2016.90

[11] Danielyan A, Katkovnik V, Egiazarian K. BM3D frames and variational image deblurring. IEEE Transactions on Image Processing. 2012;**21**(4):1715-1728. DOI: 10.1109/TIP.2011.2176954

[12] Zibulevsky M, Elad M. L1-L2 optimization in signal and image processing. IEEE Signal Processing Magazine. 2010;**27**(3):76-88. DOI: 10.1109/MSP.2010.936023

[13] Heckerman D. A Tutorial on learning bayesian networks. In: Innovations in Bayesian Networks. Berlin/Heidelberg: Springer; 2008. pp. 33-82. DOI: 10.1007/978-3-540-85066-3_3

[14] Chickering D, Heckerman D, Meek C. A Bayesian approach to learning Bayesian networks with local structure. In Proceedings of Thirteenth Conference on Uncertainty in Artificial Intelligence. 1997:80-89. Available from: https://dl.acm.org/doi/pdf/10.5555/2074226.2074236

[15] Cai J, Osher S, Shen Z. Split Bregman methods and frame based image restoration. Multiscale Modeling &

Simulation. 2009;**8**(2):337-369. DOI: 10.1137/090753504

[16] Boyd S, Parikh N, Chu E, Peleato B, Eckstein J. Distributed optimization and statistical learning via the alternating direction method of multipliers. Foundations and Trends in Machine Learning. 2011;**3**(1):1-122. DOI: 10.1561/2200000016

[17] Hoiem D, Efros A, Hebert M. Geometric context from a single image. IEEE International Conference on Computer Vision. 2005;**1**:654-661. DOI: 10.1109/ICCV.2005.107

[18] Ji H, Luo Y, Shen Z. Image recovery via geometrically structured approximation. Applied and Computational Harmonic Analysis. 2016;**41**(1):75-93. DOI: 10.1016/j.acha. 2015.08.012

[19] Simoncelli E. Bayesian denoising of visual images in the wavelet domain. Bayesian Inference in Wavelet Based Models. 1999;**141**:291-308. DOI: 10.1007/ 978-1-4612-0567-8_18

[20] Mallat S, Hwang W. Singularity detection and processing with wavelets. IEEE Transactions on Information Processing. 1992;**38**(2):617-643. DOI: 10.1109/18.119727

[21] Milanfar P. A tour of modern image filtering: New insights and methods, both practical and theoretical. IEEE Signal Processing Magazine. 2013;**30**(1): 106-128. DOI: 10.1109/MSP.2011. 2179329

[22] Sreedevi P, Hwang W, Lei S. An examplar-based approach for texture compaction synthesis and retrieval. IEEE Transactions on Image Processing. 2010; **19**(5):1307-1318. DOI: 10.1109/TIP. 2009.2039665

[23] Mairal J, Elad M, Sapiro G. Sparse representation for color image restoration. IEEE Transactions on Image Processing. 2008;**17**(1):53-69. DOI: 10.1109/TIP.2007.911828

[24] Daubechies I, Defrise M, De-Mol C. An iterative thresholding algorithm for linear inverse problems with a sparsity constraint. Communications on Pure and Applied Mathematics. 2004;**57**(11): 1413-1457. DOI: 10.1002/cpa.20042

[25] The USC-SIPI Image Database, Signal and Image Processing Institute, University of Southern California. Available from: https://sipi.usc.edu/ database/

[26] Hanif M, Seghouane A. Blind image deblurring using non-negative sparse approximation. IEEE International Conference on Image Processing. 2014. DOI: 10.1109/ICIP.2014.7025821

Section 4

Applications of the Wavelets in Waveguide Analysis and Improvement of Medical Images

Chapter 8

Application to Medical Image Processing

Anthony Y. Aidoo, Gloria A. Botchway and Matilda A.S.A. Wilson

Abstract

Medical images are often corrupted by white noise, blurring and contrast defects. Consequently, important medical information may be degraded or completely masked. Advanced medical diagnostics and pathological analysis utilize information obtained from medical images. Consequently, the best techniques must be applied to capture, compress, store, retrieve and share these images. Recently, the wavelet transform technique has been applied to enhance and compress medical images. This review focuses on the trends of wavelet-based medical image processing techniques. A summary of the application of wavelets to enhance and compress medical images such as magnetic resonance imaging (MRI), computerized tomography (CT), positron emission tomography (PET), single photon emission computed tomography (SPECT), and X-ray is provided. Morphological techniques such as closing, thinning and pruning are combined with wavelets methods to extract the features from the medical images.

Keywords:

1. Introduction

The goal of this chapter is to provide a review of the applications of wavelets to medical imaging. The focus will be on medical image denoising and compression. Advanced medical diagnostics utilize information obtained from technologies such as magnetic resonance imaging (MRI), computerized tomography (CT), positron emission tomography (PET), single photon emission computed tomography (SPECT), and X-ray [1, 2]. However, such images are corrupted by white noise, blurring and contrast defects. Consequently, important medical information may be degraded or completely masked. Recently, wavelet-based techniques have been applied to achieve superior image denoising and economical image compression.

1.1 Wavelet properties in medical imaging: multiresolution analysis

A multiresolution analysis is a decomposition of the Hilbert space $H = L^2(R)$ into a chain of closed subspaces $(V_j), j \in Z$ which form a sequence of successive approximation subspaces of H such that the following hold:

IntechOpen

1. $V_j \subset V_{j+1}$ for all $j \in Z$

2. $\bigcup_{j=-\infty}^{\infty} V_j$ is dense in $L^2(R)$ and $\bigcap_{j=-\infty}^{\infty} V_j = \{0\}$.

3. $f(x) \in V_j \Leftrightarrow f(2x) \in V_{j+1}$ for all $j \in Z$

4. $f(x) \in V_j \Leftrightarrow f(x-k) \in V_j$ for all $j, k \in Z$

5. Each subspace V_j is spanned by integer translates of a single function $f(x)$. That is, for any $f \in L^2(R)$ and any $k \in Z; f(x) \in V_0, f(x-k) \in V_0$. All subspaces are therefore scaled versions of the central space V_0.

6. (6) There exists a function $\psi(x)$, belonging to V_0, such that the sequence $(\psi(x-k); k \in Z)$ forms a Riesz basis or unconditional basis for V_0. Using the result that an orthonormal basis can always be generated out of a given Riesz basis, Riesz basis can be replaced by orthonormal basis.

1.2 Wavelet properties in medical imaging: wavelet bases

One of the special qualities of wavelets which is exploited in medical image analysis is the ability to construct L^2 bases which are simply dilations and translations of a single compactly supported function given by $\left\{\psi_{j,k} = 2^{-j/2}\psi\left(x/2^j - k\right)\right\}$ where $j, k \in Z$. This enables any image function f to be represented by:

$$f = \sum_{j \in Z} \sum_{k \in Z} c_{j,k} \psi_{j,k} \tag{1}$$

2. Undecimated wavelet transform

Conventional methods for medical image enhancement have very limited versatility in application and their use could lead to the loss of important medical image features of interest. This could be highly fatal in medical imaging applications [3]. The undecimated discrete wavelet transform (UDWT) method is a wavelet transform algorithm without the downsampling operations, resulting in both the original signal and the approximation and detailed coefficients having same length at each level of decomposition. The basic algorithm of the conventional UDWT is that it applies the transform at each point of the image and saves the detailed coefficients and uses the approximation coefficients for the next level. The size of the coefficients array does not diminish from level to level. This decomposition operation is further iterated up to a higher level. Various denoising methods using the DWT provide robust computational methods in denoising digital medical images. The only issue with the DWT is that it is shift variant. This disadvantage can may be ameriorated by using the UDWT to achieve shift invariance.

3. Image enhancement: wavelets and medical image denoising

Medical images are usually corrupted by noise inherrent in the processes of acquisition, trasmission, and retrieval [4, 5]. In particular, medical images such as those obtained from MRI or X-rays are often complicated by random noise that occurs during the image acquisition stage [6]. Wavelet-based techniques overcome most of these limitations. The objective of enhancement is to remove the effects of signal degradation caused by the signal processing. Noise may be removed by smoothing the signal by subjecting it to a low-pass filter. Sharpening to remove blur is used to identify more detailed features by applying a high-pass filter.

3.1 Wavelets methods

With its inherrent properties of multiresolution structure, application of wavelets to medical images converts the noisy image in the time domain into the wavelet transform domain. Subsequently, essential image detail information is compressed into large coefficients that are retained at different resolution scales. The small coefficients represent the noise in the image as well as any redundant information. In medical image analysis, the boundary line between “large” and “small” coefficients is crucial since it determines whether the noise is significantly removed in addition to crucial detail being preserved.

3.2 Hybrid methods for medical image denoising

Spatial filters have the tendency of blurring images since the technique smoothens data in order to remove niose [7]. Tackling this problem by relying on wavelet transform alone sometimes does not satisfactorily address the image enhancement problem since wavelet tranform methods are plagued by oscillations, shift variance, aliasing and lack of directionality. Three methods that are combined with wavelets significantly eliminate the problems listed above the medical image enhancement outcomes considered here.

3.2.1 Total variation denoising

Total variation (TV) regularization is a deterministic method that minimizes the effect of discontinuities in image processing [8]. The TV technique is endowed with the power of preserving and even enhancing the edges. The use of TV for image denoising assumes that the observed image is made up of the sum of a piecewise smooth image and guassian noise.

A real valued function $f(x)$, representing a signal is sampled using the partition $P = \{-\infty < x_0 < x_1 < \dots < x_n, n \in \mathbb{N}\}$ of the interval $[x_0, x]$. The TV T_f of f over the interval is defined by:

$$T_f(x) = \sup\left\{\sum_{i}^{n} |f(x_i) - f(x_{i-1})| : -\infty < x_0 < x_1 < \dots < x_n, n \in \mathbb{N}\right\} \tag{2}$$

If $\lim_{x\to\infty} T_f(x)$ is finite, then f is of bounded variation. The TV of an L^1 function f of several variables, in an open subset Ω of $\mathbb{R}^n$, is defined as by:

$$T_f(x) = \sup\left\{\int_{\Omega} f(x)\mathrm{div}\phi(x)\mathrm{d}x : \phi(x) \in C_c^1(\Omega, \mathbb{R}^n), \|\phi\|_{L^\infty(\Omega)} \le 1\right\} \tag{3}$$

If f is a differentiable function defined on a bounded open domain $\Omega \subset \mathbb{R}^n$ this reduces to:

$$T_f(x) = \int_\Omega |\nabla f(x)| \mathrm{d}x \tag{4}$$

Definition 3.1 *The total variation of an image is defined by the duality: for $u \in L^1_{loc}$ the total variation is given by $T_f = \sup\{-\mathrm{div}\phi \mathrm{d}x : \phi \in C_c^\infty(\Omega; \mathbb{R}^N), |\phi(x)| \leq 1 \forall x \in \Omega\}$.*

This hybrid approach used here represents a noisy image in a simplified form by Eq. (1). The reconstruction of $u(x)$ reduces to the optimization problem of minimizing the function

$$E(u) = \frac{\lambda}{2} \|u - z\|^2_{L^2(\Omega)} + R(u) \tag{5}$$

(see for example [9]). Here, the parameter $\lambda > 0$ and $R(u)$ is the regularization functional defined on the domain Ω. The disadvantage of this method is that despite removing noise adequately, it removes essential details from the image [8]. Since the efficiency of the method is controlled by the choice of the regularization functional, this is usually costly in medical imaging. The use of the TV of the image function below ameliorates this problem.

$$R(u) = T_z(u) = \int_\Omega |\Delta u| dx \tag{6}$$

It leads to sharper reconstruction of the original image by both removing the imbedded noise and better preservation of its edges [10, 11]. TV minimization scheme takes the geometric information of the original images into account, and this helps to preserve and sharpen the edges significantly [11].

3.2.2 The wavelet-total variation method

Proposition 1 *[12] Let $K = \{p \in L^2(\Omega) : \int_\Omega p(x)u(x)\mathrm{d}x \leq T_z(u) \forall u \in L^2(\Omega)\}$. If T_z is considered as a functional over the Hilbert space $L^2(\Omega)$, we have $\partial T_z(u) = \{p \in K : \int_\Omega p(x)u(x)\mathrm{d}x = T_z(u)\}$.*

Proof 1 *If $p \in K$ and $\int_\Omega p(x)u(x)\mathrm{d}x = T_z(u)$ then $p \in \partial T_z(u)$. Clearly for any $v \in L^2(\Omega)$ we have $T_z(v) = \sup_{p \in K} \int_\Omega p(x)u(x)\mathrm{d}x$. $T_z(v) \geq \int_\Omega p(x)v(x)\mathrm{d}x = T_z(u) + \int_\Omega (v(x) - u(x))p(x)\mathrm{d}x$. Conversely, if $p \in \partial T_z(u)$, then for any $t > 0$ and $v \in \mathbb{R}^N$, with $T_z(tu) = tT_z(u)$ since T_z is positively one-homogeneous, we have: $tT(v) = T_z(v) \geq T_z(u) + \int_\Omega p(x)(tv(x) - u(x))\mathrm{d}x$. Dividing by t and letting $t \to \infty \mapsto$ leads to $T_z(v) \int_\Omega p(x)v(x)\mathrm{d}x$. Hence $p \in K$. On the other hand, letting $t \to 0$ gives $T_z(u) \leq \int_\Omega p(x)u(x)\mathrm{d}x$.*

Method	Sensitivity	Specificity	Accuracy	PSNR
TV	79.0%	80.0%	91.7%	30.97%
TV and UDWT	82.5%	93.3%	97.0%	42.80%

Table 1.
TV vs TV and UDWT.

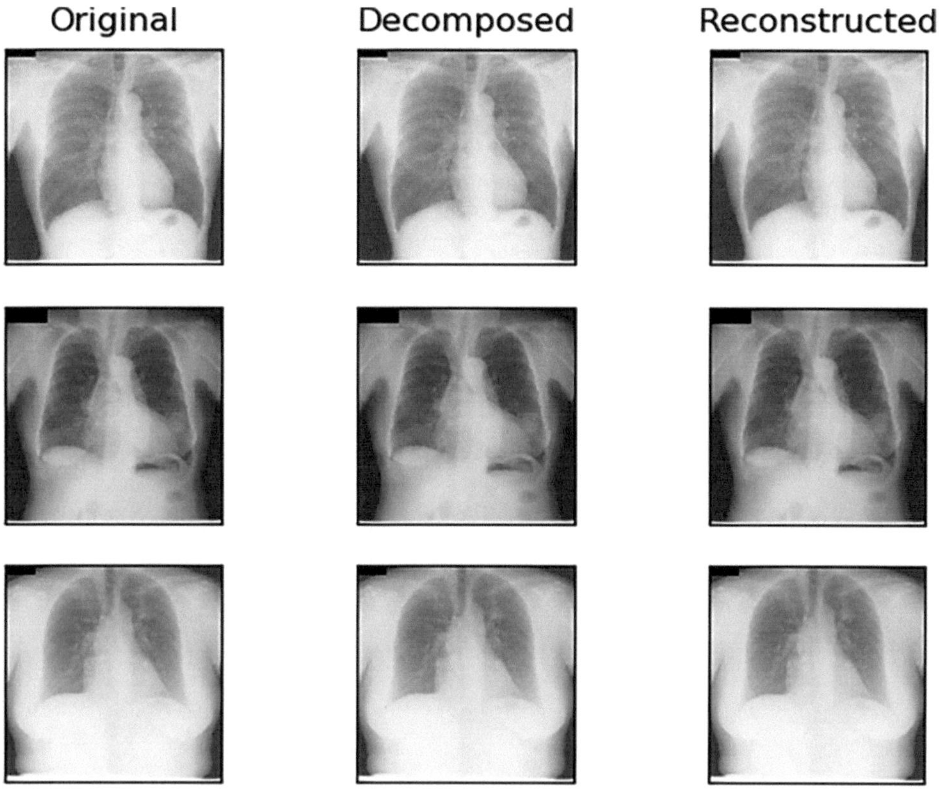

Figure 1.
(Column 1) Original chest images with nodules, (column 2) wavelet decomposition of images in column 1, (column 3) reconstructed images from the decomposed images.

The wavelet TV scheme represents the components of the function by orthogonal wavelet basis. The wavelet coefficients are then selected to achieve the goals of denoising and enhancement (**Table 1**). **Figure 1** shows the results on the chest radiograph images.

The Python 2.7 code is given below.

```
"""A code to implement a wavelet denoising and morphological enhancement
"""
#import math
#import numpy as np

import cv2
import mat      ot as plt import pywt
kernel=cv2.getStructuringElement(cv2.MORPH_ELLIPSE,(11,11))
img1=cv2.imread('JPCLN001.jpg') coeff1=pywt.wavedec2(img1, 'bior1.3')
coeff11=pywt.waverec2(coeff1,'db2')
erosion1=cv2.erode(coeff11,kernel,iterations=2)
opening1=cv2.dil te(erosion1,kernel,iterations=3)
oinv1=1-opening1
fig 1=plt.figure()
fig 1.suptitle('Original, Decomposed and Reconstructed CR Images with Nodules')
```

```
plt.subplot(331),plt.imshow(img1),plt.title('a'),plt.xticks([]),plt.yt icks([])
plt.subplot(332),plt.imshow(coeff1[0]),plt.title('d'),plt.xticks([]),p lt.yticks([])
plt.subplot 333),plt.imshow(coeff11),plt.title('g'),plt.xticks([]),plt .yticks([])
fig 1.savefig('1CR i ages with nodules, decomposed and reconstructed.png')
fig 3=plt.figure()
fig 3.suptitle('Decomposed nodule images eroded, opened and inversed')
plt.subplot(331),plt.imshow(erosion1),plt.title('a'),plt.xticks([]),pl t.yticks([])
plt.subplot(332),plt.imshow(opening1),plt.title('d'),plt.xticks([]),pl t.yticks([])
plt.subpl t(333),plt.imshow(oinv1),plt.title('g'),plt.xticks([]),plt.y
ticks([])
fig 3.savefig('1Decomposed nodule images eroded, opened and inversed.png')
```

3.2.3 Mathematical morphology

Mathematical morphology (MM) is a technique for extracting image components of interest Dilation and erosion are the two basic operations in mathematical mophology as well as thinning, opening, closing, and prunning. Wavelets combined with MM has recently been used to improve chest radiographs [13].

Definition 3.2 *Erosion and Dilation: Let E be the Euclidean space, let* $A : E \subseteq \mathbb{Z}^2 \to \mathbb{Z}$ *be an image and* $B : \mathbb{Z}^2 \to \{0, 1\}$ *be a structuring element. The translation of a set C by a point* $z = (z_1, z_2)$*, denoted by* $(C)_z$ *is defined as* $(C)_z = a|a = c + z, c \in A$*. The erosion of A by B, denoted by* $(A \ominus B)$*, is expressed as*

$$(A \ominus B) = z|(B)_z \subseteq A, \tag{7}$$

ie. The set of all pixel locations z in the image plane where $(B)_z$ is contained in A.

Definition 3.3 *The dilation of A by B is denoted by* $(A \oplus B)$ *and is expressed as*

$$(A \oplus B) = z|(\hat{B})_z \cap A \neq \varnothing, \tag{8}$$

where $\hat{B} = w|w = -b, \text{for } b \in B$ *is the reflection of* B.

This indicates the set of all pixel locations z in the image plane where the intersection of $\hat{B}$ with A is not empty [14].

Erosion shrinks an image or a region A by a template or a structuring element B. Dilation expands an image or a region A by a template or a structuring element B. The dilation process consists of obtaining the reflection of B about its origin and then shifting this reflection by some displacement x.

Other effects can be obtained by applying erosion and dilation in a loop. Closing and opening are two examples of basic erosion and dilation combinations.

3.3 Opening and closing

Definition 3.4 *The opening of A by B, denoted by* $A \circ B$*, is simply erosion of A by B, followed by dilation of the result by B, that is,*

$$(A \circ B) = (A \ominus B) \oplus B. \tag{9}$$

Visually, opening smoothens contours, breaks narrow isthmuses and eliminates small islands.

Definition 3.5 *The closing of A by B, denoted by $A \bullet B$, is a dilation followed by an erosion and is given as*

$$(A \bullet B) = (A \oplus B) \ominus B. \tag{10}$$

Closing smoothens the contours, fills narrow gulfs and eliminates small holes. It is based on these operations that other operations are derived.

3.4 Thinning and pruning

The thinning operation is related to the hit-or-miss transform and it can be expressed in terms of it. The thinning operation is derived by translating the origin of the structuring element to each possible pixel position in the image and comparing it with the underlying image pixels at each such position. Pruning is a post-processing technique that follows thinning. It removes parasitic components known as spurs which are unwanted branches, from the thinned image. There are specific structuring elements used for pruning.

Combined with MM wavelets can be used to decompose a fingerprint image in order to extract the areas with details. The results of this approach is shown in **Figures 2** and **3**.

Algorithm 1
Denoising the fingerprint image 1. Load fingerprint image. 2. Convert the greyscale fingerprint image into a binary image. 3. Decompose image using wavelets into detailed and approximated parts. 4. Reconstruct the fingerprint image using the detailed parts of the decomposed image and set the approximated part to zero.

Algorithm 2
Processing the image for feature extraction 1. Load denoised image. 2. Perform a binary closing on the image to close all insignificant holes in the image. 3. Thin image. 4. Prune the result to remove spurs and spikes. 5. Extract features.

3.4.1 Wavelet K-SVD approach

The wavelet tranform technique (for image denoising) has several advantages such as sparsity, multiresolution structure, and similarity with human vision. Recently, it has been combined with K-Singular Valued Decomposition (K-SVD) algorithm, and an adaptive learning over the wavelet decomposition of a noisy medical image has resulted [15].

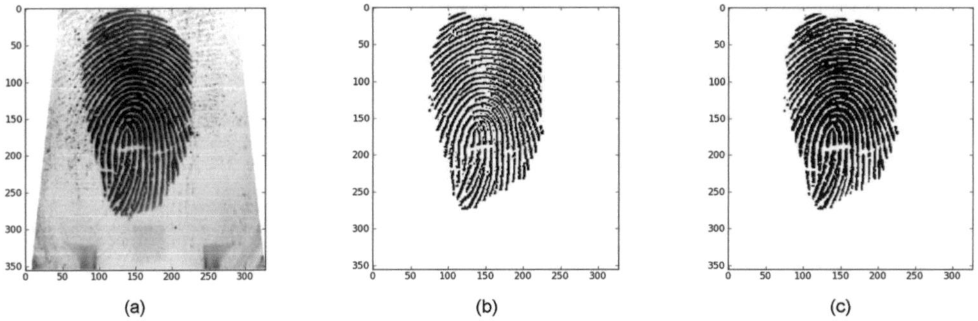

Figure 2.
Wavelet analysis and synthesis of image: (a) original image, (b) decomposed image, (c) reconstructed image.

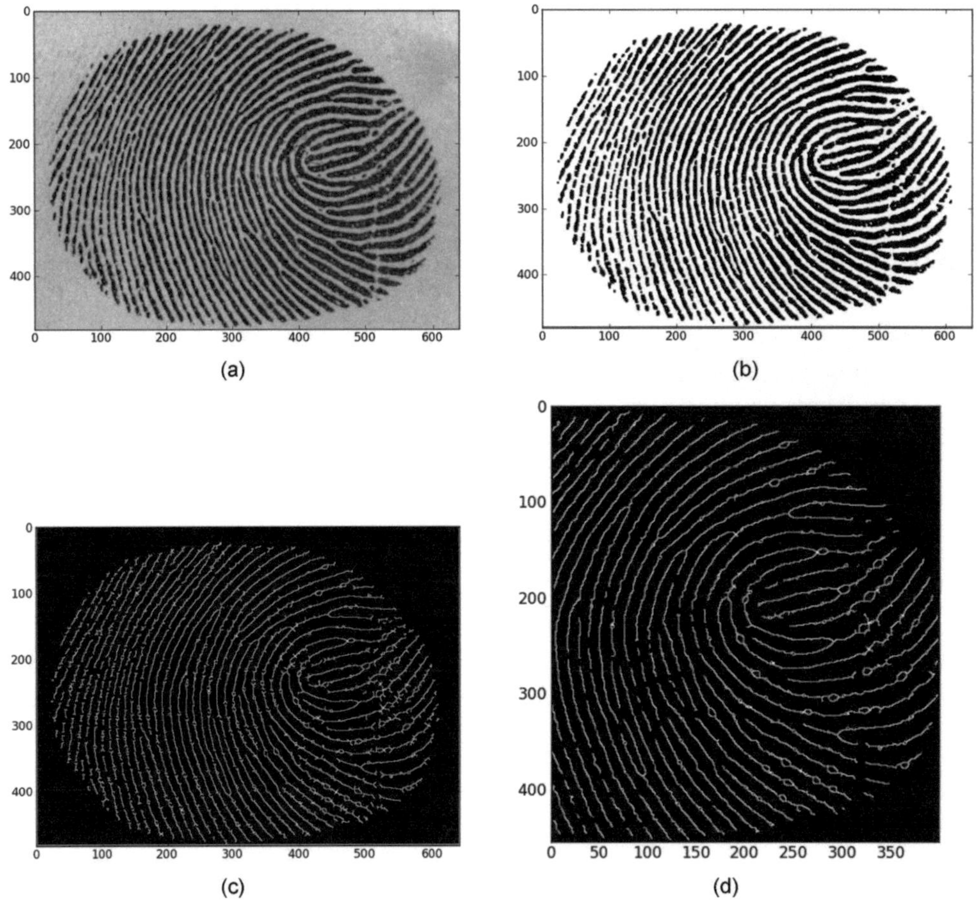

Figure 3.
Morphological operations applied to fingerprint images: (a) original image, (b) binary closing of image, (c) thinned image, (d) prunned image.

4. Wavelet based compression methods for storing medical images

In medical image compression, a mathematical transform is applied to the digital data. This is intended to compress the data for efficient storage, transmission, and

Method	DWT (Haar) [17]	DWT (dB) [17]	DWT and MM
CR	7.40%	21.86%	21.90%

Table 2.
Compression ratios.

retrieval. Compression involves coding and image approximation and helps to reduce the quantity of information, improve the transmission rate and reduce the size of the equipment and storage space required. Data compression requires the choice of a transform such that as many of the transformed data as possible vanish. The ability to localize basis functions in wavelet applications make them suitable for compression. In addition to this property of wavelets, a wavelet decomposition of an image capitalizes on its multiresolution structure, a recursive method to compute the wavelet transform of an image [16]. The three components of the wavelet tranform based image are image decomposition, quantization, and decompression.

4.1 Hybrid methods for medical image compression

The DWT combined with vector quantization methods have recently be shown to achieve superior results in medical image compression than wavelet alone technique. For example, Ammah and Owusu [17] proposed an efficient medical hybrid image decompression technique for ultrasound and MRI images. The method consists of first preprocessing the image by noise removal. This is follwed by filtering the image using the DWT with hard thresholding. The image is subsequently vector quantized and then Huffman encoded. The inverse operations are then applied to obtain the decompressed image.

The metrics used to evaluate the efficiency of the compression methods are the compression ratio (CR) and the peak signal to noise ratio (PSNR) given by:

$$\mathrm{CR} = \frac{I(x,y)}{I'(x,y)} \tag{11}$$

and

$$\mathrm{PSNR} = 20\log_{10}\left\{\frac{255}{\mathrm{RMSE}}\right\} \tag{12}$$

where the RMSE is the root mean squared error.

Using a new class of spline wavelet filters, more effective data compression techniques have been devised to compress massive quantities of medical image data leading to a more economical storage process and enhanced medical image quality when retrieved. Combined with other methods, the inherrent properties of wavelets such as sparsity and multiresolution structure produce superior medical image data compresssion results than the competition. CRs for three most used methods are is shown in **Table 2**.

5. Conclusion

Modern radiology techniques are essential in advanced medical diagnostics and pathological analysis [18]. Applications of the wavelet tranform for medical imaging techniques and current advances in research in this direction has been highlighted in this chapter. This includes the combination of the standard wavelet techniques with TV, MM and other methods.

Acknowledgements

Author Anthony Y. Aidoo acknowledges the support from the CSU-AAUP Faculty Research Fund 2021-22.

Classifications

2020 AMS Subject Classification: 68U10, 94A12, 28A99, 92C55, 65R10, 44A99.

Author details

Anthony Y. Aidoo[1*], Gloria A. Botchway[2] and Matilda A.S.A. Wilson[3]

1 Department of Mathematical Sciences, Eastern Connecticut State University, Willimantic, USA

2 Department of Mathematics, University of Ghana, Accra, Ghana

3 Department of Computer Science, University of Ghana, Accra, Ghana

*Address all correspondence to: aidooa@easternct.edu

References

[1] Georgieva V, Petrov P, Zlatareva D. Medical image processing based on multidimensional wavelet transforms—Advantages and trends. AIP Conference Proceedings. 2021;**2333**:1-7

[2] Ferroukhi M, Ouahabi A, Attari M, Habchi Y, Taleb-Ahmed A. Medical video coding based on 2nd-generation wavelets: Performance evaluation. Electronics. 2019;**8**:1-18. DOI: 10.3390/electronics8010088

[3] Unaldi N, Temel S, Demerci S. Undecimated wavelet transform based contrast enhancement. International Journal of Computer, Electrical, Automation, Control and Information Engineering. 2013;**7**:1215-1218

[4] Patil R, Bhosale S. Medical image denoising using wavelet transform and singular value decomposition. WEENTECH Proceedings in Energy. 2019;**6**(2):1-8

[5] Chervyakov N, Lyakhov P, Nagornov N. Analysis of the quantization noise in discrete wavelet transform filters for 3D medical imaging. Applied Sciences. 2020;**10**(4):1223. DOI: 10.3390/app10041223

[6] Quahabi A. Image denoising using wavelets: Application in medical imaging. Advances in Heuristic Signal Processing and Applications. Berlin, Heidelberg: Springer; 2013. pp. 287-313. DOI: 10.1007/978-3-642-37880-5-13

[7] Raj P, Venkateswarlu T. Denoising medical images using dual tree complex wavelet transform. Procedia Technology. 2012;**4**:238-244. DOI: 10.1016/j.protcy.2012.05.036

[8] Estrela VV, Megalhaes HA, Saotome O. Total variation applications in computer vision. In: Handbook of Research on Emerging Perspectives in Intelligent Pattern Recognition, Analysis, and Image Processing. Hershey, Pennsylvania: IGI Global; 2016. pp. 41-64

[9] Wilson M, Hafron JB, Acquah AY, Aidoo. A total variation-undecimated wavelet approach to chest radiograph image enhancement. TELKOMNIKA. 2019;**17**(4):2116-2124

[10] Vese L. A study in the bounded variation space of denoising-deblurring variational problem. Applied Mathematics and Optimization. 2001; **44**(2):131-161. DOI: 10.1007/s00245-001-0017-7

[11] Starck J-L, Elad M, Donoho D. Redundant multiscale transforms and their applications for morphological component analysis. Advances in Imaging and Electron Physics. 2004;**132**

[12] Chambolle A, Caselles V, Novaga M, Cremers D, Pock T. An introduction to total variation for image analysis. Theoretical Foundations and Numerical Methods for Sparse Recovery. 2010;**9**: 263-340. DOI: 10.1515/9783110226157.263

[13] Aidoo AY, Wilson M, Botchway GA. Chest radiograph image enhancement with wavelet. TELKOMNIKA. 2019; **17**(5):2587-2594. DOI: 10.12928/TELKOMNIKA.v17i5.11964

[14] Bansal R, Sehgal P, Bedi P. Effective morphological extraction of true fingerprint minutiae based on the hit or miss transform. International Journal of Biometrics and Bioinformatics. 2020; **4**(2):71. DOI: 10.1.1.170.3632

[15] Bnou K, Raghay S, Hakim A. A wavelet denoising approach based on

unsupervised learning model. EURASIP Journal on Advances in Signal Processing. 2020;**36**:1-26

[16] Schomer DF, Elekes AA, Hazle J, Huffman JC, Thompson SK, Chui CK, et al. Introduction to wavelet-based compression of medical images. Imaging and Therapeutic Technology. 1998; **18**(2):469-481

[17] Ammah PNT, Owusu E. Robust medical image compression based on wavelet transform and vector quantization. Informatics in Medicine Unlocked. 2019;**15**:1-11

[18] Khan SA, Khan A, Song O, Nazir M. Medical imaging fusion techniques: A survey benchmark analysis, open challenges and recommendations. Journal of Medical Imaging and Health Informatics. 2020;**10**(11):2523-2531. DOI: 10.1166/jmihi.2020.3222

Chapter 9

Straight Rectangular Waveguide for Circular Dielectric Material in the Cross Section and for Complementary Shape of the Cross Section

Zion Menachem

Abstract

This chapter presents wave propagation along a straight rectangular waveguide for practical applications where there are two complementary shapes of the dielectric profile in the cross section. In the first case, the cross section consists of circular dielectric material in the center of the cross section. In the second case, the cross section consists of a circular hollow core in the center of the cross section. These examples show two discontinuous cross sections and complementary shapes that cannot be solved by analytical methods. We will explain in detail the special technique for calculating the dielectric profile for all cases. The method is based on Laplace and Fourier transforms and inverse Laplace and Fourier transform. In order to solve any inhomogeneous problem in the cross section, more than one technique can be proposed for the same mode-model method. We will explain in detail how and where the technique can be integrated into the proposed mode-model. The image method and periodic replication are needed for fulfilling the boundary condition of the metallic waveguide. The applications are useful for straight rectangular waveguides in millimeter regimes, where the circular dielectric material is located in the center of the cross section, and also for hollow waveguides, where the circular hollow core is located in the center of the cross section.

Keywords: wave propagation, dielectric profiles, rectangular waveguide, circular dielectric material, circular hollow core

1. Introduction

We begin with a review of numerical and approximate methods for the modal analysis of general optical dielectric waveguides with emphasis on recent developments as published in [1]. Six groups of methods were reviewed: the finite-element method, the finite-difference method, the integral-equation method, methods based on series expansion, approximate methods based on separation of variables, and methods that do not fit the above groups.

The use of wavelet-like basis functions for solving electromagnetics problems is demonstrated in [2]. The modes of an arbitrarily shaped hollow metallic waveguide use a surface integral equation and the method of moments. A class of wavelet-like basis functions produces a sparse method of moments. A technique for efficient computation of an integral wavelet transform of a finite-energy function on a dense set of the time-scale domain is proposed [3] by using compactly supported spline wavelets. Application of principal component analysis and wavelet transform to fatigue crack detection in waveguides is proposed in [4]. Ultrasonic guided waves are a useful tool in structural health monitoring applications that can benefit from built-in transduction, moderately large inspection ranges, and high sensitivity to small flaws. An accurate full-wave integral formulation was developed [5] for the study of integrated planar dielectric waveguide structures with printed metalized sections, which are of practical interest for millimeter-wave and submillimeter-wave applications. An advantageous finite element method for the rectangular waveguide problem was developed [6] by which complex propagation characteristics may be obtained for arbitrarily shaped waveguides. The finite-element method has been used to derive approximate values of the possible propagation constant for each frequency. The impedance characteristics of the fundamental mode in a rectangular waveguide were computed using this finite element method. The extension to higher-order elements is straightforward, and by modifications of the method it is possible to treat other types of waveguides as well, e.g., dielectric waveguides with impedance walls and open unbounded dielectric waveguides properties treating the region of infinity.

A comprehensive study of the design and performance of a multilayer dielectric rod waveguide with a rectangular cross section is proposed in [7]. The design is comprised of a high permittivity core encased by a low permittivity cladding. A mathematical model was proposed to predict the fundamental mode cutoff frequency in terms of the core dimensions and the core and cladding permittivity. The model is useful for design purposes and it offers an excellent match to full-wave electromagnetic simulation results.

The characteristics of the effective-medium-clad dielectric waveguides, including dispersion, cross-polarization, crosstalk between parallel waveguides, bending loss, and wave leakage at the crossing, have been comprehensively investigated and measured [8].

Mode matching has been done at all the air and dielectric interfaces and thus the characteristic equations have been derived [9]. Two ratios are introduced in the characteristic equations and the new set of characteristic equations thus obtained are then plotted and graphical solutions are obtained for the propagation parameters assuming certain numerical values for the introduced ratios.

A fundamental and accurate technique to compute the propagation constant of waves in a dielectric rectangular waveguide was proposed [10]. The formulation is based on matching the fields to the constitutive properties of the material at the boundary.

The method of lines for the analysis of dielectric waveguides was proposed [11]. These waveguides are uniform along the direction of propagation, are loss-free and passive. Hybrid-mode dispersion curves, field and intensity distributions for integrated optical waveguides were presented.

The problem of normal waves in a closed regular waveguide of arbitrary cross section has been considered [12]. It was reduced to a boundary value problem for the longitudinal components of the electromagnetic field in Sobolev spaces. The solutions were defined using the variational formulation of the problem. The problem was reduced to the study of an operator function. The properties of the operators involved in the operator function were examined. Theorems were proved concerning the discrete character of the spectrum and the distribution of characteristic numbers of the operator function on the complex plane. The completeness of the system of Eigen- and associated vectors of the operator function was investigated.

TE-wave propagation in a hollow waveguide with a graded dielectric layer using a hyperbolic tangent function is proposed in Ref. [13]. General formulas for the electric field components of the TE-waves, applicable to hollow waveguides with arbitrary cross sectional shapes were presented. The exact analytical results for the electric field components were illustrated in the special case of a rectangular waveguide. The exact analytical results for the reflection and transmission coefficients are valid for waveguides of arbitrary cross sectional shapes. The obtained reflection and transmission coefficients are in exact asymptotic agreement with those obtained for a very thin homogeneous dielectric layer using mode-matching and cascading. The proposed method gives analytical results that are directly applicable without the need of mode-matching, and it has the ability to model realistic, smooth transitions.

Rectangular waveguides were the earliest mode of transmission lines used for compact systems like radars and inside equipment shelters [14]. An air-filled rectangular waveguide WR-90 is simulated using HFSS simulation software to obtain different parameters. The electric and magnetic field patterns are analyzed: intrinsic impedance and wavelength for the first four modes of the waveguide are also obtained.

The diffraction of electromagnetic waves by rectangular waveguides with a longitudinal slit has been simulated [15]. The results allow determining the patterns of change in frequency bands in which the structure can be used as a directional coupler and as a power divider when changing the number of slots, their sizes and provisions. Modeling the characteristics of such kinds of structures allows predicting the creation of directional couplers and power dividers with high integral characteristics.

Several methods of propagation along the straight waveguides were developed, based on Maxwell's equations. A transfer matrix function for the analysis of electromagnetic wave propagation along the straight dielectric waveguide with arbitrary profiles has been proposed in Ref. [16].

In this chapter, the main objective is to generalize the mode model method [16] in order to solve also complicated and practical problems of circular dielectric material and a circular hollow core in the center of the cross section of the rectangular waveguide. It is important to distinguish between the mode-model method and the proposed technique. The proposed technique deals only with calculating the dielectric profile in the cross section of the inhomogeneous case. In order to solve any inhomogeneous problem in the cross section, more than one technique can be proposed for the same mode-model method. The technique proposed in this chapter will refer to two interesting practical applications. In the first case, the cross section consists of circular dielectric material in the center of the cross section. In the second case, the cross section shows the complementary shape of the cross section of the first case, as an example of a hollow waveguide in which the circular hollow core is located in the center of the cross section. These examples show two discontinuous cross sections and complementary shapes that cannot be solved by analytical methods. We will explain in detail the special technique for calculating the dielectric profile in all cases. After receiving the expressions of the proposed technique for each inhomogeneous problem in the cross section, we will explain how and where the technique can be integrated into the proposed mode-model. The second objective is to find the relevant parameters in order to obtain the Gaussian behavior of the output field in the interesting cases of circular dielectic material and a circular hollow core in the rectangular cross section.

2. Complementary shapes in the cross section for different applications

The wavelet transform creates a representation of the signal in both the time and frequency domain in order to allow efficient access to localized information about the signal. A set of waveforms comprising a transform is called a basis function. Fourier transforms use only sine and cosine waves as their basis functions, namely a signal is decomposed into a series sine and cosine functions or wavelets by the FFT. Examples for the applications of wavelet transform are demonstrated in [2–5]. The proposed method in this chapter is based on the Fourier transform that creates a representation of the signal in the frequency domain. Two complicated and complementary shapes are given in this section.

Figure 1(a) and **(b)** shows two complementary shapes of profiles in the cross section of the straight rectangular waveguide and their relevant parameters. The circular dielectric material in the center of the cross section is shown in **Figure 1(a)** and the circular hollow core in the center of the cross section is shown in **Figure 1(b)**. The two examples are demonstrated as a response to a half-sine (TE_{10}) input-wave profile. These two different complementary shapes of the cross section are demonstrated for two different applications. The first example (**Figure 1(a)**) is useful in millimeter regimes where the circular dielectric material is located in the center of the cross section. The second example (**Figure 1(b)**) is useful in the millimeter regimes where the circular hollow core is located in the center of the cross section.

The main objective is to generalize the mode model method [16] in order to also solve complicated problems of circular dielectric material and a circular hollow core in the rectangular cross section. All the mathematical development relates to the frequency domain. The main points are given in Appendix A.

It is important to separate between mode-model method and the proposed technique. The proposed technique refers only to calculating of the dielectric profile in the

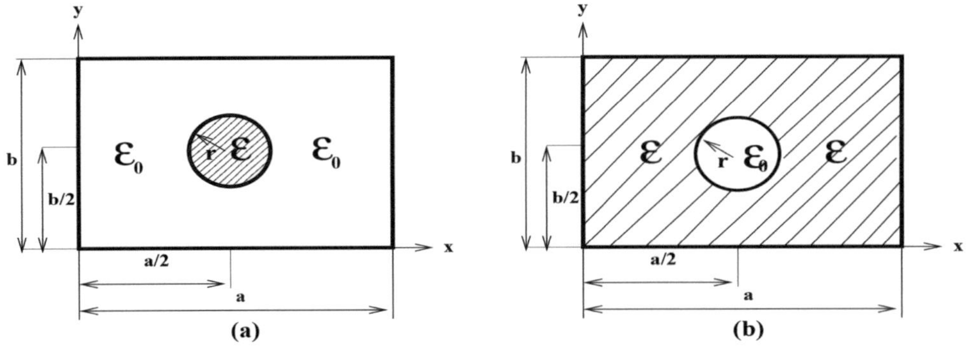

Figure 1.
Complementary shapes of profiles in the cross section of the straight rectangular waveguide and their relevant parameters. (a) Circular dielectric material in the center of the cross section. (b) A circular hollow core in the center of the cross section.

cross section of the inhomogeneous problem. In order to solve any inhomogeneous problem in the cross section, more than one technique can be proposed for the same mode-model method. After receiving the expressions of the proposed technique for eace inhomogeneous problem in the cross section, we will explain how and where the technique can be integrated into the proposed mode-model. The second objective is to find the relevant parameters in order to obtain the Gaussian behavior of the output field in the interesting cases of circular dielectic profile and circular hollow profile in the rectangular cross section.

The method is based on Maxwell's equations for the computation of output fields at each point along the straight waveguide. This method relates the wave profile at the output to the input wave in the Laplace space. A Laplace transform is necessary to obtain convenient and simple input–output connections of the fields. The method consists of Fourier coefficients of the transverse dielectric profile and of the input–output profile. Thus, the accuracy of the method depends on the number of the modes in the system.

The output transverse field profiles are computed by the inverse Laplace and Fourier transforms. The output components of the electric field are given finally by

$$\mathbf{E_x} = \{\mathbf{D_x} + \alpha_1\mathbf{M_1}\mathbf{M_2}\}^{-1}(\hat{\mathbf{E}}_{\mathbf{x_0}} - \alpha_2\mathbf{M_1}\hat{\mathbf{E}}_{\mathbf{y_0}}), \tag{1}$$

$$\mathbf{E_y} = \{\mathbf{D_y} + \alpha_1\mathbf{M_3}\mathbf{M_4}\}^{-1}(\hat{\mathbf{E}}_{\mathbf{y_0}} - \alpha_3\mathbf{M_3}\hat{\mathbf{E}}_{\mathbf{x_0}}), \tag{2}$$

$$\mathbf{E_z} = \mathbf{D_z^{-1}}\left\{\hat{\mathbf{E}}_{\mathbf{z_0}} + \frac{\mathbf{1}}{\mathbf{2s}}(\mathbf{G_x}\mathbf{E}_{\mathbf{x_0}} + \mathbf{G_y}\mathbf{E}_{\mathbf{y_0}}) - \frac{1}{2}(\mathbf{G_x}\mathbf{E_x} + \mathbf{G}_y\mathbf{E_y})\right\}, \tag{3}$$

where $\mathbf{E_{x_0}}, \mathbf{E_{y_0}}, \mathbf{E_{z_0}}$ are the initial values of the corresponding fields at z = 0, i.e., $\mathbf{E_{x_0}} = \mathbf{E_x}$ (x, y, z = 0), and $\hat{\mathbf{E}}_{\mathbf{x_0}}, \hat{\mathbf{E}}_{\mathbf{y_0}}, \hat{\mathbf{E}}_{\mathbf{z_0}}$ are the initial-value vectors.

The modified wave-number matrices are given by

$$\mathbf{D_x} \equiv \mathbf{K}^{(0)} + \frac{k_o^2\chi_0}{2s}\mathbf{G} + \frac{jk_{ox}}{2s}\mathbf{N}\mathbf{G_x}, \qquad \mathbf{D_y} \equiv \mathbf{K}^{(0)} + \frac{k_o^2\chi_0}{2s}\mathbf{G} + \frac{jk_{oy}}{2s}\mathbf{M}\mathbf{G_y}, \qquad \mathbf{D_z} \equiv \mathbf{K}^{(0)} + \frac{k_o^2\chi_0}{2s}\mathbf{G}, \tag{4}$$

where the diagonal matrices $\mathbf{K}^{(0)}$, $\mathbf{M}$, and $\mathbf{N}$ are given by

$$\mathbf{K}^{(\mathbf{0})}_{(\mathbf{n},\mathbf{m})(\mathbf{n}',\mathbf{m}')} = \left\{\left[\mathbf{k_o^2} - (\mathbf{n}\pi/\mathbf{a})^2 - (\mathbf{m}\pi/\mathbf{b})^2 + \mathbf{s}^2\right]/\mathbf{2s}\right\}\delta_{\mathbf{nn}'}\delta_{\mathbf{mm}'}, \qquad \mathbf{M}_{(\mathbf{n},\mathbf{m})(\mathbf{n}',\mathbf{m}')} = \mathbf{m}\delta_{\mathbf{nn}'}\delta_{\mathbf{mm}'}, \qquad \mathbf{N}_{(\mathbf{n},\mathbf{m})(\mathbf{n}',\mathbf{m}')} = \mathbf{n}\delta_{\mathbf{nn}'}\delta_{\mathbf{mm}'}, \tag{5}$$

and where

$$\alpha_1 = \frac{k_{ox}k_{oy}}{4s^2}, \quad \alpha_2 = \frac{jk_{ox}}{2s}, \quad \alpha_3 = \frac{jk_{oy}}{2s}, \qquad \mathbf{M_1} = \mathbf{N}\mathbf{G_y}\mathbf{D_y^{-1}}, \quad \mathbf{M_2} = \mathbf{M}\mathbf{G_x}, \quad \mathbf{M_3} = \mathbf{M}\mathbf{G_x}\mathbf{D_x^{-1}}, \quad \mathbf{M_4} = \mathbf{N}\mathbf{G_y}. \tag{6}$$

Similarly, the other components of the magnetic field are obtained. The output transverse field profiles are given by the inverse Laplace and Fourier transforms, as follows

$$\mathbf{E_y}(\mathbf{x}, \mathbf{y}, \mathbf{z}) = \sum_{\mathbf{n}} \sum_{\mathbf{m}} \int_{\sigma - \mathbf{j}_\infty}^{\sigma + \mathbf{j}_\infty} \mathbf{E_y}(\mathbf{n}, \mathbf{m}, \mathbf{s}) \exp\left[\mathbf{jnk_{ox}x} + \mathbf{jmk_{oy}y} + \mathbf{sz}\right] \mathbf{ds}, \tag{7}$$

where the inverse Laplace transform is calculated according to the Salzer method [17, 18]. The inverse Laplace transform is performed in this study by a direct numerical integration on the Laplace transform domain by using the method of Gaussian Quadrature. The integration path in the right side of the Laplace transform domain includes all the singularities.

$$\int_{\sigma - j_\infty}^{\sigma + j_\infty} e^{s\zeta} E_y(s) ds = \frac{1}{\zeta} \int_{\sigma - j_\infty}^{\sigma + j_\infty} e^p E_y(p/\zeta) dp = \frac{1}{\zeta} \sum_{i=1}^{15} w_i E_y(s = p_i/\zeta), \tag{8}$$

where w_i and p_i are the weights and zeros, respectively, of the orthogonal polynomials of order 15. The Laplace variable s is normalized by p_i/ζ in the integration points, where $\mathrm{Re}(p_i) > 0$ and all the poles should be localized on their left side on the Laplace transform domain. This approach of a direct integral transform does not require as in other methods, to deal with each singularity separately.

The relation between the functions f (t) and F (p) is given by

$$f(t) = \frac{1}{2\pi j} \int_{\sigma - j_\infty}^{\sigma + j_\infty} e^{pt} F(p) dp. \tag{9}$$

The function F(p) may be either known only numerically or too complicated for evaluating f(t) by Cauchy's theorem. The function F(p) behaves like a Polynomial without a constant term, in the variable 1/p, along $(\sigma - j_\infty, \sigma + j_\infty)$. One may find f(t) numerically by using new quadrature formulas (analogous to those employing the zeros of the Laguerre polynomials in the direct Laplace transform). A suitable choice of p_i yields an n-point quadrature formula that is exact when $p_2 n$ is any arbitrary polynomial of the 2n(th) degree in $x \equiv 1/p$, namely

$$\frac{1}{2\pi j} \int_{\sigma - j_\infty}^{\sigma + j_\infty} e^p \rho(1/p) dp = \sum_{i=1}^{n} A_i(n) \rho_{2n}(1/p_i). \tag{10}$$

In Eq. (10), $x_i \equiv 1/p_i$ are the zeros of the orthogonal polynomials $p_n(x) \equiv \Pi(x - x_i)$ where

$$\frac{1}{2\pi j} \int_{\sigma - j_\infty}^{\sigma + j_\infty} e^p \left(\frac{1}{p}\right) p_n \left(\frac{1}{p}\right) \left(\frac{1}{p}\right)^i dp = 0, \tag{11}$$

i = 0,1, … ,$n - 1$ and $A_i(n)$ correspond to the Christoffel numbers. The normalization $P_n(1/p) \equiv (4n - 2)(4n - 6), \ldots, 6p_n(1/p)$, for $n \geq 2$, produces all integral coefficients. $P_n(1/p)$ is proven to be $(-1)^n e^{-p} p^n d^n (e^p/p^n)/dp^n$. The numerical table gives us the values of the reciprocals of the zeros of $P_n(x)$ or $p_i(n)$, the zeros of $P_n(x)$, or $1/p_i{}^n$, and the corresponding Christoffel numbers $A_i(n)$. By using these quantities in the quadrature formula that represents in Eq. (10), then the "Christoffel numbers" are given by

$$A_i(n) \equiv \frac{1}{2\pi j} \int_{\sigma - j_\infty}^{\sigma + j_\infty} e^p L_i(n+1) \left(\frac{1}{p}\right) dp. \tag{12}$$

A sufficient condition for Eq. (12) to hold is obviously the "Orthogonality" of $(1/p)p_n(1/p)$ with respect to any "arbitrary" $\rho(1/p)$ (see Eq. (11)). The points $1/p_i$ are denoted by $1/(p_i)^n$ and they are the "zeros" of a certain set of "orthogonal polynomials" in the variable $1/p$. By using these quantities in the "quadrature formula" we can obtain theoretically "exact accuracy" for "any polynomial" in $1/p$ up to the 16(th) degree.

A Fortran code is developed using NAG subroutines (The Numerical Algorithms Group (NAG)) [19].

The proposed technique will introduce details for all the interesting cases of a discontinuous cross section, as shown in **Figure 1(a)** and **(b)**.

3. Calculation of the different inhomogeneous profiles

This section explains the proposed technique for calculating the dielectric profile for the two different inhomogeneous and complicated shapes of the cross section, as shown in **Figure 1(a)** and **(b)**.

3.1 Calculation for circular dielectric material in the center of the cross section

The technique is based on Fourier transform and uses the image method and periodic replication for fulfilling the boundary conditions of the metallic waveguide. Periodicity and symmetry properties are chosen to force the boundary conditions at the location of the walls in a real problem, by extending the waveguide region ($0 \le x \le a$, and $0 \le y \le b$) to regions that are four-fold larger ($-a \le x \le a$, and $-b \le y \le b$). The elements of the matrix g(n,m) are calculated for an arbitrary profile in the cross section of the straight waveguide according to **Figure 2 (a)** and **(b)**.

The dielectric profile $g(x,y)$ is calculated according to $\varepsilon(x,y) = \varepsilon_0(1+g(x,y))$ and according to **Figure 2(a)** and **(b)** where $g(x,y) = g_0$. The specific case of circular dielectric material in the center of the cross section is shown in **Figure 2(b)** by using the image method. We obtain

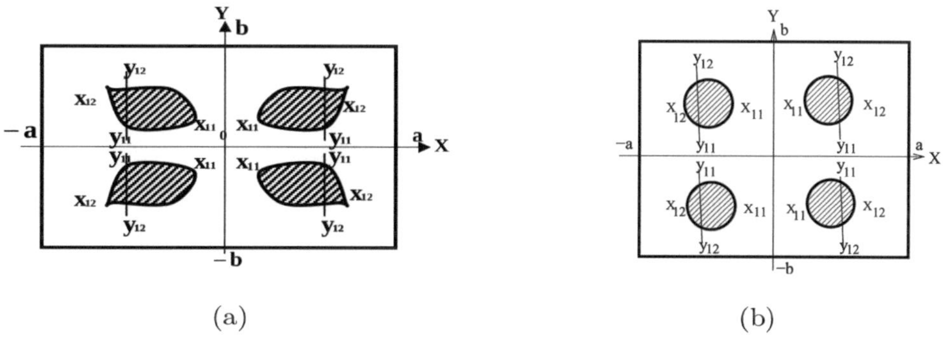

Figure 2.
The image method for (a) an arbitrary profile in the cross section, and (b) the specific case of circular dielectric material in the center of the cross section.

$$
\begin{aligned}
g(n,m) &= \frac{g_0}{4ab}\int_{-a}^{a} dx\int_{-b}^{b} \exp\left[-j\left(k_x x + k_y y\right)\right]dy \\
&= \frac{g_0}{4ab}\left\{\int_{x_{11}}^{x_{12}} dx\int_{y_{11}}^{y_{12}} \exp\left[-j\left(k_x x + k_y y\right)\right]dy + \int_{-x_{12}}^{-x_{11}} dx\int_{y_{11}}^{y_{12}} \exp\left[-j\left(k_x x + k_y y\right)\right]dy\right. \\
&\left. + \int_{-x_{12}}^{-x_{11}} dx\int_{-y_{12}}^{-y_{11}} \exp\left[-j\left(k_x x + k_y y\right)\right]dy + \int_{x_{11}}^{x_{12}} dx\int_{-y_{12}}^{-y_{11}} \exp\left[-j\left(k_x x + k_y y\right)\right]dy\right\} \\
&= \frac{g_0}{4ab}\left\{\int_{x_{11}}^{x_{12}} dx\int_{y_{11}}^{y_{12}} \exp\left[-j\left(k_x x + k_y y\right)\right]dy + \int_{x_{12}}^{x_{11}} -dx\int_{y_{11}}^{y_{12}} \exp\left[-j\left(-k_x x + k_y y\right)\right]dy\right. \\
&\left. + \int_{x_{12}}^{x_{11}} -dx\int_{y_{12}}^{y_{11}} - \exp\left[-j\left(-k_x x - k_y y\right)\right]dy + \int_{x_{11}}^{x_{12}} dx\int_{y_{12}}^{y_{11}} - \exp\left[-j\left(k_x x - k_y y\right)\right]dy\right\} \\
&= \frac{g_0}{4ab}\left\{\int_{x_{11}}^{x_{12}} \exp\left[j(k_x x)\right]dx\int_{y_{11}}^{y_{12}} \exp\left[j\left(k_y y\right)\right]dy + \int_{x_{11}}^{x_{12}} \exp\left[j(k_x x)\right]dx\int_{y_{11}}^{y_{12}} \exp\left[-j\left(k_y y\right)\right]dy\right. \\
&\left. + \int_{x_{11}}^{x_{12}} \exp\left[-j(k_x x)\right]dx\int_{y_{11}}^{y_{12}} \exp\left[j\left(k_y y\right)\right]dy + \int_{x_{11}}^{x_{12}} \exp\left[-j(k_x x)\right]dx\int_{y_{11}}^{y_{12}} \exp\left[-j\left(k_y y\right)\right]dy\right\} \\
&= \frac{g_0}{4ab}\left\{\int_{x_{11}}^{x_{12}} dx\int_{y_{11}}^{y_{12}} \exp\left[-j\left(k_x x + k_y y\right)\right]dy + \int_{x_{11}}^{x_{12}} dx\int_{y_{11}}^{y_{12}} \exp\left[j\left(k_x x - k_y y\right)\right]dy\right. \\
&\left. + \int_{x_{11}}^{x_{12}} dx\int_{y_{11}}^{y_{12}} \exp\left[j\left(k_x x + k_y y\right)\right]dy + \int_{x_{11}}^{x_{12}} dx\int_{y_{11}}^{y_{12}} \exp\left[-j\left(k_x x - k_y y\right)\right]dy\right\} \\
&= \frac{g_0}{2ab}\left\{\int_{x_{11}}^{x_{12}} \left(\exp\left(jk_x x\right) + \exp\left(-jk_x x\right)\right)dx\int_{y_{11}}^{y_{12}} \cos\left(k_y y\right)dy\right\}.
\end{aligned}
\tag{13}
$$

If y_{11} and y_{12} are functions of x, then we obtain

$$
\begin{aligned}
g(n,m) &= \frac{g_0}{abk_y}\int_{x_{11}}^{x_{12}} \left[\sin\left(k_y y_{12}(x)\right) - \sin\left(k_y y_{11}(x)\right)\right]\cos\left(k_x x\right)dx \\
&= \frac{2g_0}{am\pi}\int_{x_{11}}^{x_{12}} \sin\left[\frac{m\pi}{2b}\left(y_{12}(x) - y_{11}(x)\right)\right]\cos\left[\frac{m\pi}{2b}\left(y_{12}(x) + y_{11}(x)\right)\right]\cos\left(\frac{n\pi}{a}x\right)dx,
\end{aligned}
\tag{14}
$$

where $k_x = (n\pi)/a$, and $k_y = (m\pi)/b$.

The radius of the circle is given by $r = \sqrt{(x - a/2)^2 + (y - b/2)^2}$, thus for the specific case of the cross section (**Figure 1(a)**) and according to the image method, we obtain.

$$y_{11}(x) = b/2 - \sqrt{r^2 - (x - a/2)^2}, \tag{15}$$

$$y_{12}(x) = b/2 + \sqrt{r^2 - (x - a/2)^2} \tag{16}$$

The dielectric profile for the cross section (**Figure 1(a)**) is given by.

$$g(n,m\neq 0)=\frac{2g_0}{am\pi}\int_{x_{11}}^{x_{12}}\sin\left[\frac{m\pi}{2b}\left(y_{12}(x)-y_{11}(x)\right)\right]\cos\left[\frac{m\pi}{2b}\left(y_{12}(x)+y_{11}(x)\right)\right]\cos\left(\frac{n\pi}{a}x\right)dx, \tag{17}$$

$$g(n,m=0)=\frac{g_0}{ab}\int_{x_{11}}^{x_{12}}\left(y_{12}(x)-y_{11}(x)\right)\cos\left(\frac{n\pi}{a}x\right)dx, \tag{18}$$

where $y_{12}(x)-y_{11}(x)=2\sqrt{r^2-(x-a/2)^2}$ and $y_{12}(x)+y_{11}(x)=b$.

The cyclic matrix **G** is given as follows. The Fourier transform is applied to the transverse dimension

$$\bar{g}(k_x,k_y)=F\{g(x,y)\}=\int_x\int_y g(x,y)e^{-jk_xx-jk_yy}dxdy. \tag{19}$$

The components are organized in a vectorial notation as follows

$$\mathbf{E}=\begin{bmatrix}\overline{E}_{-N,-M}\\ \vdots\\ \overline{E}_{-N,+M}\\ \vdots\\ \overline{E}_{+n,+m}\\ \vdots\\ \overline{E}_{+N,+M}\end{bmatrix}. \tag{20}$$

The Fourier components of the dielectric profile are calculated in the Fourier space. The convolution operation

$$\bar{g}^*\overline{E}=\left\{\sum_{n'=-N}^{N}\sum_{m'=-M}^{M}g_{n-n',m-m'}E_{n',m'}\right\} \tag{21}$$

is written in a matrix form as $\mathbf{G}E$ where

$$\bar{g}(n,m)(n',m')=g_{n-n',m-m'} \tag{22}$$

and the matrix order is (2 N + 1)(2 M + 1), where E is the electric field.

The convolution operation is expressed by the cyclic matrix **G** which consists of Fourier components of the dielectric profile $\bar{g}_{nm}$. Thus, the cyclic matrix **G** is given by the form

$$\mathbf{G}=\begin{bmatrix} g_{00} & g_{-10} & g_{-20} & \cdots & g_{-nm} & \cdots & g_{-NM}\\ g_{10} & g_{00} & g_{-10} & \cdots & g_{-(n-1)m} & \cdots & g_{-(N-1)M}\\ g_{20} & g_{10} & \ddots & \ddots & \ddots & & \\ \vdots & g_{20} & \ddots & \ddots & \ddots & & \\ g_{nm} & \ddots & \ddots & \ddots & g_{00} & \vdots & \\ \vdots & & & & & & \\ g_{NM} & \cdots & \cdots & \cdots & \cdots & \cdots & g_{00}\end{bmatrix}. \tag{23}$$

The derivative of the dielectric profile is given by

$$g_x(n,m) = \frac{2g_0}{am\pi}\int_{x_{11}}^{x_{12}} \sin\left[\frac{m\pi}{2b}\left(y_{12}(x) - y_{11}(x)\right)\right] \cos\left[\frac{m\pi}{2b}\left(y_{12}(x) + y_{11}(x)\right)\right] \cos\left(\frac{n\pi}{a}x\right)dx, \tag{24}$$

where y_{11} and y_{12} are given according to Eqs (15) and (16). Similarly, we can calculate the value of $g_y(n,m)$, where $g_y(x,y) = (1/\varepsilon(x,y))(d\varepsilon(x,y)/dy)$.

3.2 Calculation for the circular hollow core in the center of the cross section

Figure 3(a)–(c) shows the extending of the waveguide region in all cases to a four-fold larger region, according to the image method. The image method and periodic replication are needed for fulfilling the boundary condition of the metallic waveguide. **Figure 3(a)** shows the hollow waveguide where the circular hollow core is located in the center of the cross section. This figure represents an example of the complementary shape of **Figure 3(c)**. **Figure 3(b)** shows the cross section entirely filled with the dielectric material. **Figure 3(c)** shows the cross section where the circular dielectric material is located in the center.

Note that the problem shown in **Figure 3(a)** is more complicated than the problem shown in **Figure 3(c)**, and the technique for solving this inhomogeneous problem in the cross section based on the image method is not effective for the specific case shown in **Figure 3(a)**. Thus the proposed technique for calculating the dielectric profile of this problem is based on the fact that this figure represents an example of the complementary shape of **Figure 3(c)**.

In order to solve any inhomogeneous problem in the cross section (e.g., **Figure 3 (a)** and **(c)**), more than one technique can be proposed for the same mode-model method.

The proposed technique to calculate the dielectric profile for the cross section as shown in **Figure 3(a)** for hollow waveguide is based on subtracting the dielectric profile of the waveguide with the dielectric material in the core (**Figure 3(c)**) from the dielectric profile of the waveguide filled entirely with the dielectric material (**Figure 3(b)**).

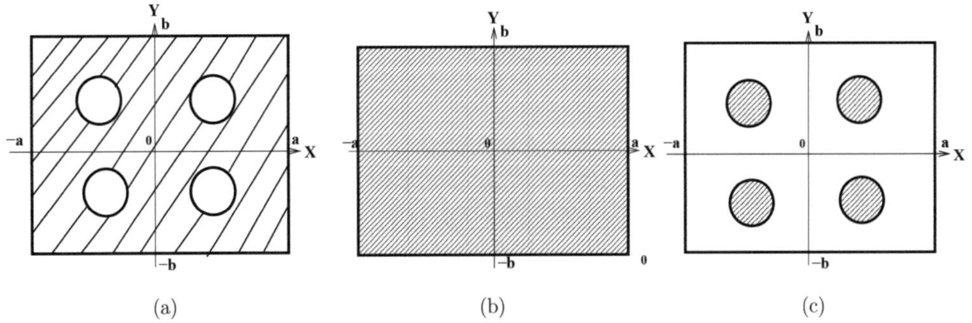

Figure 3.
Extending the waveguide region in all cases to a four-fold larger region, according to the image method. (a) The hollow waveguide where the circular hollow core is located in the center of the cross section. (b) The cross section entirely filled with dielectric material. (c) Circular dielectric material is located in the center of the cross section.

4. Numerical results

This section presents several examples for the different geometries of two specific examples of the complementary shapes of dielectric profile in the cross section, as shown in **Figure 1(a) and (b)**. The solutions are demonstrated as a response to a half-sine (TE_{10}) input-wave profile.

A comparison with the known transcendental Equation [20] according to **Figure 4(a)** is given in order to examine the validity of the theoretical model. The known solution for the dielectric slab modes based on the transcendental Equation [20] is given as follows:

$$E_{y1} = j\frac{k_z}{\varepsilon_0}\sin(\nu x) \qquad 0 < x < t \tag{25}$$

$$E_{y2} = j\frac{k_z}{\varepsilon_0}\frac{\sin(\nu t)}{\cos(\mu(t-a/2))}\cos[\mu(x-a/2)] \qquad t < x < t+d \tag{26}$$

$$E_{y3} = j\frac{k_z}{\varepsilon_0}\sin[\nu(a-x)] \qquad t+d < x < a, \tag{27}$$

where $\nu \equiv \sqrt{k_o^2 - k_z^2}$ and $\mu \equiv \sqrt{\varepsilon_r k_o^2 - k_z^2}$ result from the transcendental equation

$$\left(\frac{a-d}{d}\right)\frac{d\mu}{2}\tan\left(\frac{d\mu}{2}\right) - (t\nu)\cot(t\nu) = 0. \tag{28}$$

The solution obtained for the wave profile ((25)–(27)) describes a symmetrical mode of the dielectric slab. This mode is substituted as an input wave at z = 0 to the solution of the proposed theoretical model Eq. (2).

The comparison between the theoretical model (Eq. (2)) and the transcendental equation (Eqs (25)–(27)) is shown in **Figure 4(b)** for the dielectric slab in a rectangular metallic waveguide (**Figure 4(a)**) and the convergence of our theoretical results is shown in **Figure 4(c)**.

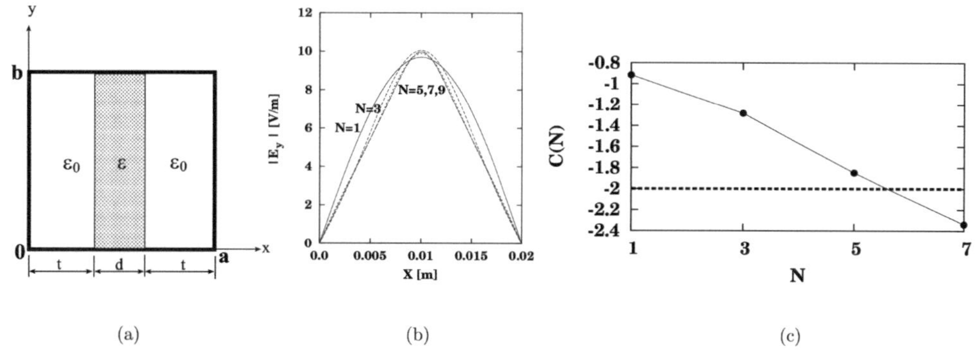

Figure 4.
(a) A dielectric slab in a rectangular metallic waveguide. (b) A comparison between the theoretical model (Eq. (2)) and the transcendental equation (Eqs (25)–(27)) according to Ref. [20], where a = 2b = 2 cm, d = 3.3 mm, ε_r= 9, and λ= 6.9 cm. (c) The convergence of our theoretical results.

The comparison is demonstrated for every order ($N = 1, 3, 5, 7$, and 9). The order N determines the accuracy of the solution. The convergence of the solution is verified by the criterion for the E_y component of the fields.

The convergence of the solution is verified by the criterion

$$C(N) \equiv log \left\{ \frac{\max\left(|E_y^{N+2} - E_y^N|\right)}{|\max\left(E_y^{N+2}\right) - \min\left(E_y^N\right)|} \right\}, \qquad N \geq 1. \qquad (29)$$

where the number of the modes is equal to $(2N+1)^2$. The order N determines the accuracy of the solution.

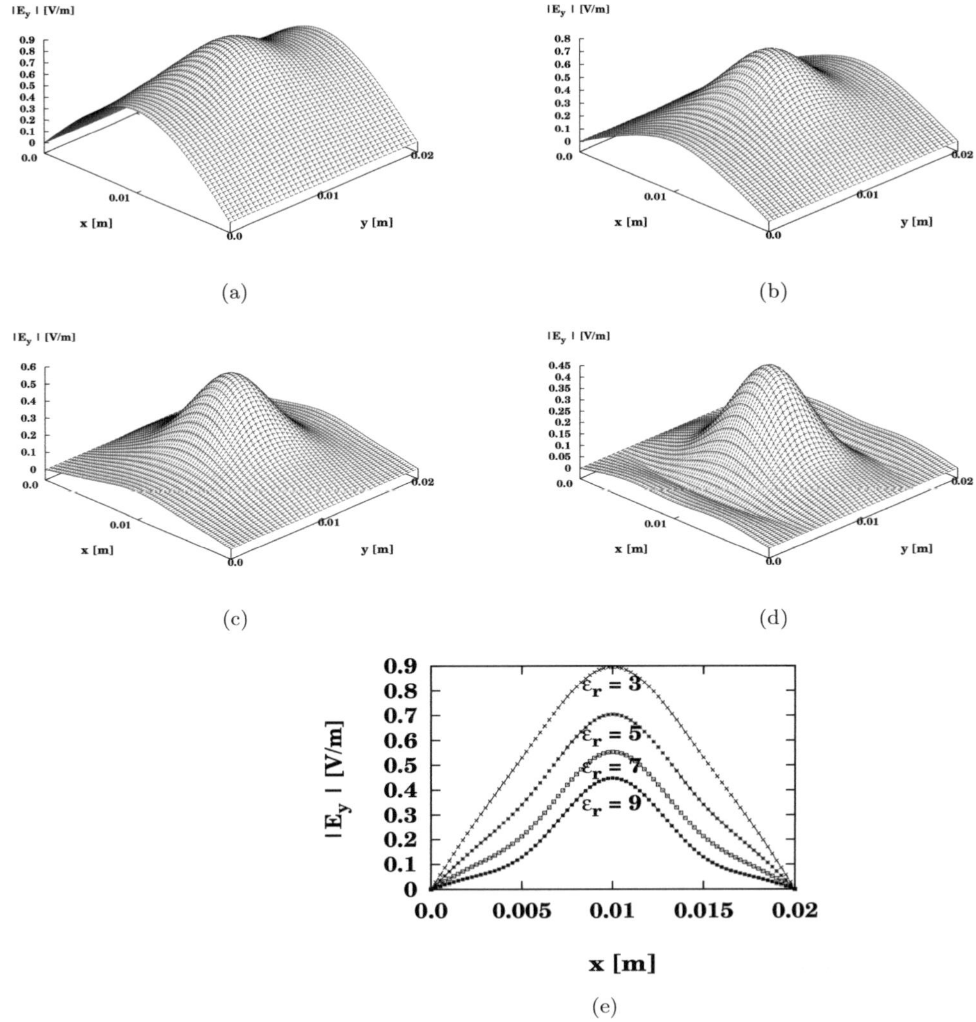

Figure 5.
The output field where the circular dielectric material is located in the center of the cross section of the straight rectangular waveguide, where a = b = 20 mm, and r = 2.5 mm, r is the radius of the circular dielectric material, and for (a) ε_r = 3, for (b) ε_r = 5, for (c) ε_r = 7, and for (d) ε_r = 9. (e) The output field in the same cross section of the results (a)–(d) for x-axis and where y = b/2 = 10 mm, for the values of ε_r = 3, 5, 7, and 9, respectively. The other parameters are a = b = 20 mm, k_o = 167 1/m, λ = 3.75 cm, and β =58 1/m.

If the value of the criterion (Eq. (29)) is less than -2, then the numerical solution is well converged. When N increases, then $E_y(N)$ approaches E_y. The value of the criterion between N = 7 and N = 9 is equal to $-2.38 \simeq -2$, namely a hundredth part. Comparison between the theoretical mode-model (Eq. (2)) and the known model [20] shows good agreement.

Figure 5(a)–(e) shows the output field where the circular dielectric material is located in the center of the cross section of the straight rectangular waveguide, where a = b = 20 mm, ε_r = 3, 5, 7, 9, for r = 2.5 mm, where r is the radius of the circular dielectric material. The output field in the same cross section of the results **Figure 5 (a)–(d)** are shown in **Figure 5(e)** for the x-axis and where y = b/2 = 10 mm, for the

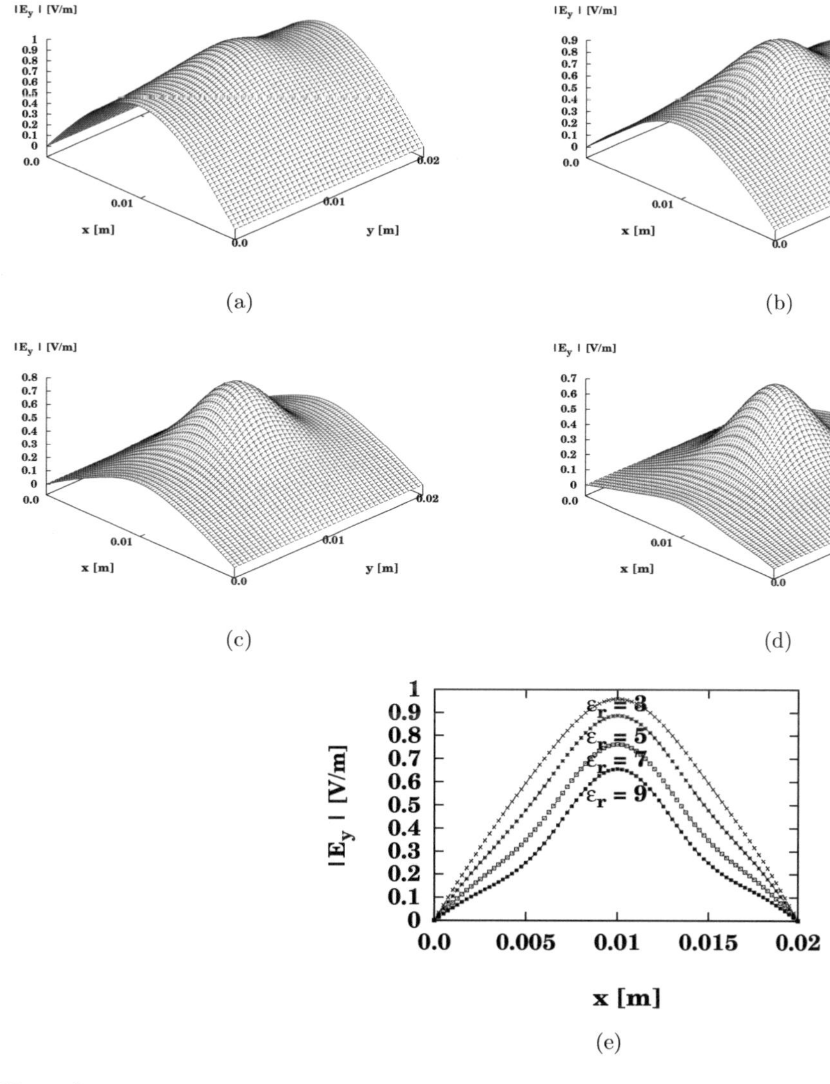

Figure 6.
The output field where the circular dielectric material is located in the center of the cross section of the straight rectangular waveguide, where a = b = 20 mm, and r = 2 mm, where r is the radius of the circular dielectric material, and for (a) ε_r = 3, for (b) ε_r = 5, for (c) ε_r = 7, and for (d) ε_r = 9. (e) The output field in the same cross section of the results (a)–(d) for x-axis and where y = b/2 = 10 mm, for the values of ε_r = 3, 5, 7, and 9, respectively. The other parameters are a = b = 20 mm, k_o = 167 1/m, λ = 3.75 cm, and β = 58 1/m.

values of ε_r = 3, 5, 7, and 9, respectively. **Figure 6(a)–(e)** demonstrates the output fields by changing only the parameter of the radius of the circular dielectric material from r = 2.5 to r = 2. The other parameters are a = b = 20 mm, k_0 = 167 1/m, λ = 3.75 cm, and β = 58 1/m.

Figure 7(a)–(e) shows the output field where the circular hollow core is located in the center of the cross section of the straight rectangular waveguide, where a = b = 20 mm, ε_r = 1.5, 1.6, 1.7, 1.8, for r = 2.5 mm, where r is the radius of the circular hollow core. The output field in the same cross section of the results **Figure 7 (a)–(d)** are shown in **Figure 7(e)** for x-axis and where y = b/2 = 10 mm, for the values of ε_r = 1.5, 1.6, 1.7, and 1.8, respectively. **Figure 8(a)–(e)** demonstrates the output

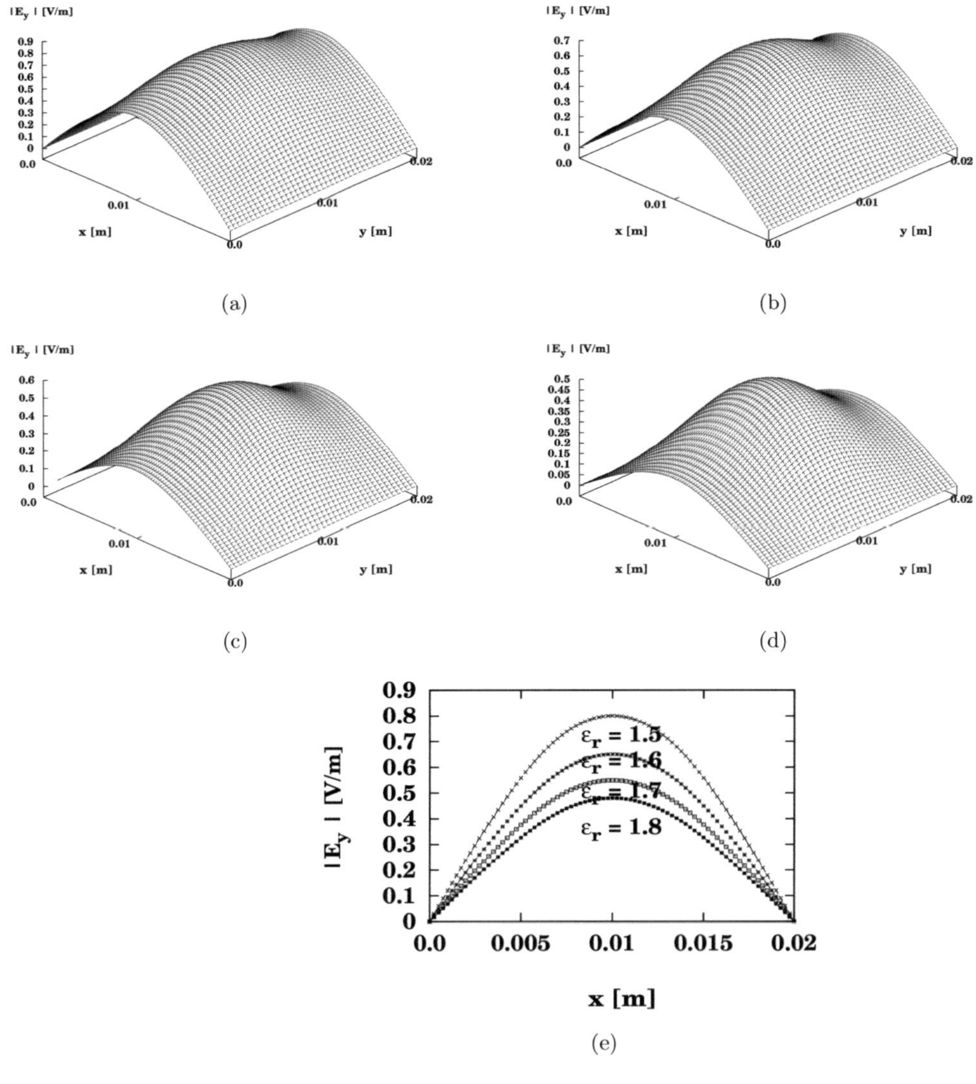

Figure 7.
The output field where the circular hollow core is located in the center of the cross section of the straight rectangular waveguide, where a = b = 20 mm, and r = 2.5 mm, where r is the radius of the circular hollow core, and for (a) ε_r = 1.5, for (b) ε_r = 1.6, for (c) ε_r = 1.7, and for (d) ε_r = 1.8. (e) The output field in the same cross section of the results (a)–(d) for x-axis and where y = b/2 = 10 mm, for the values of ε_r = 1.5, 1.6, 1.7, and 1.8, respectively. The other parameters are a = b = 20 mm, k_o = 167 1/m, λ = 3.75 cm, and β = 58 1/m.

fields by changing only the parameter of the radius of the circular hollow core from r = 2.5 to r = 2. The other parameters are a = b = 20 mm, k_0 = 167 1/m, λ = 3.75 cm, and β = 58 1/m.

By increasing only the dielectric constant from $\varepsilon_r = 3$ to $\varepsilon_r = 9$, according to **Figures 5(a)–(e)** and **6(a)–(e)**, and from $\varepsilon_r = 1.5$ to $\varepsilon_r = 1.8$, according to **Figures 7(a)–(e)** and **8(a)–(e)**, the Gaussian shape of the output transverse profile of the field increased, the TE_{10} wave profile decreased, and the relative amplitude of the output field decreased.

We can predict the waveguide parameters (ε_r and r) for obtaining the Gaussian behavior of the output field in all case. The cross section in the first interesting

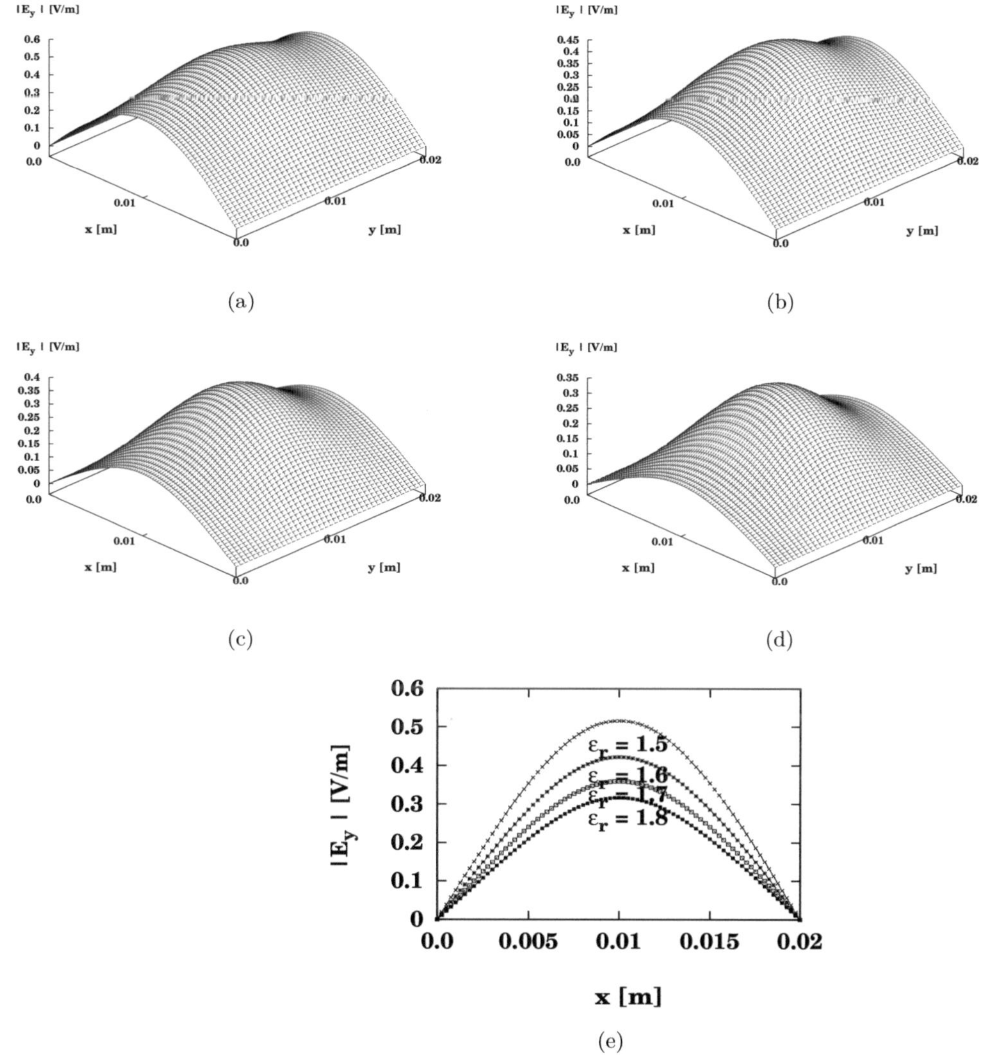

Figure 8.
The output field where the circular hollow core is located in the center of the cross section of the straight rectangular waveguide, where a = b = 20 mm, and r = 2 mm, where r is the radius of the circular hollow core, and for (a) ε_r = 1.5, for (b) ε_r = 1.6, for (c) ε_r = 1.7, and for (d) ε_r = 1.8. (e) The output field in the same cross section of the results (a)–(d) for x-axis and where y = b/2 = 10 mm, for the values of ε_r = 1.5, 1.6, 1.7, and 1.8, respectively. The other parameters are a = b = 20 mm, k_o = 167 1/m, λ = 3.75 cm, and β =58 1/m.

case consists of circular dielectric material in the center of the cross section (**Figure 1(a)**). The cross section in the second interesting case consists of a circular hollow core in the center of the cross section (**Figure 1(b)**). The output results refer to the same parameters a = b = 20 mm, k_0 = 167 1/m, λ = 3.75 cm, and β = 58 1/m. According to the results of the first case, in order to obtain the Gaussian behavior, the values of ε_r = 3, 5, 7, 9 and r = 2 or r = 2.5 are needed. In the second case, in order to obtain the Gaussian behavior, the values of ε_r = 1.5, 1.6, 1.7, and 1.8 and r = 2 or r = 2.5 are needed.

These results are strongly affected by the different parameters ε_r and r, and for the same other parameters of k_0 = 167 1/m, λ = 3.75 cm, β = 58 1/m, and the dimensions of the rectangular cross section.

5. Conclusions

The wavelet transform creates a representation of the signal in both the time and frequency domain in order to allow efficient access of localized information about the signal. A set of waveforms comprising a transform is called a basis function. Fourier transforms use only sine and cosine waves as their basic functions, namely a signal is decomposed into a series of sine and cosine functions or wavelets by the FFT. Examples for the applications of wavelet transform are demonstrated in [2–5]. The proposed method in this chapter is based on the Fourier transform that creates a representation of the signal in the frequency domain.

Two specific examples of complementary shapes of dielectric profile in the cross section were introduced in this chapter. In the first case, the cross section consists of circular dielectric material in the center of the cross section. In the second case, the cross section shows the complementary shape of the cross section of the first case, as an example of a hollow waveguide in which the circular hollow core is located in the center of the cross section.

Note that the problem shown in **Figure 3(a)** is more complicated than the problem shown in **Figure 3(c)**, and the technique for solving this inhomogeneous problem in the cross section based on the image method is not effective for the specific case shown in **Figure 3(a)**. The proposed technique for calculating the dielectric profile of the problem shown in **Figure 3(a)** is based on the fact that this figure represents an example of the complementary shape of **Figure 3(c)**.

In order to solve any inhomogeneous problem in the cross section (e.g., **Figure 3(a)** and **(c)**), more than one technique can be proposed for the same mode-model method.

The proposed technique to calculate the dielectric profile for the cross section as shown in **Figure 3(a)** for hollow waveguide is based on subtracting the dielectric profile of the waveguide from the dielectric material in the core (**Figure 3(c)**) from the dielectric profile of the waveguide filled entirely with the dielectric material (**Figure 3(b)**).

Figures 5(a)–(e) and **6(a)–(e)** demonstrate the output fields, where the circular dielectric material is located in the center of the cross section of the straight rectangular waveguide, where the parameter r refers to the radius of the circular dielectric material. **Figures** 7(a)–(e) and **8(a)–(e)** demonstrate the output fields, where the circular hollow core is located in the center of the cross section of the straight rectangular waveguide, where the parameter r refers to the radius of the circular hollow core. The other parameters are a = b = 20 mm, k_0 = 167 1/m, λ = 3.75 cm, and β = 58 1/m.

By increasing only the dielectric constant from $\varepsilon_r = 3$ to $\varepsilon_r = 9$, according to **Figures 5(a)–(e)** and **6(a)–(e)**, and from $\varepsilon_r = 1.5$ to $\varepsilon_r = 1.8$, according to **Figures 7(a)–(e)** and **8(a)–(e)**, the Gaussian shape of the output transverse profile of the field increased, the TE_{10} wave profile decreased, and the relative amplitude of the output field decreased.

We can predict the waveguide parameters (ε_r and r) for obtaining the Gaussian behavior of the output field in all cases. The output results refer to the same parameters a = b = 20 mm, $k_0 = 167$ 1/m, $\lambda = 3.75$ cm, and $\beta = 58$ 1/m. According to the results of the first case, in order to obtain the Gaussian behavior, the values of $\varepsilon_r = 3, 5, 7, 9$ and r = 2 or r = 2.5 are needed. In the second case, in order to obtain the Gaussian behavior, the values of $\varepsilon_r = 1.5, 1.6, 1.7$, and 1.8 and r = 2 or r = 2.5 are needed.

The results are strongly affected by the different parameters ε_r and r, and for the same other parameters of $k_0 = 167$ 1/m, $\lambda = 3.75$ cm, $\beta = 58$ 1/m, and the dimensions of the rectangular cross section.

The applications are useful for straight rectangular waveguides in millimeter regimes, where the circular dielectric material is located in the center of the cross section, and also for hollow waveguides, where the circular hollow core is located in the center of the cross section.

Appendix A

The wavelet transform creates a representation of the signal in both the time and frequency domain in order to allow efficient access of localized information about the signal. A set of waveforms comprising a transform is called a basis function. Fourier transforms use only sine and cosine waves as its basic functions, namely a signal is decomposed into a series sine and cosine functions or wavelets by the FFT. Examples for the applications of wavelet transform are demonstrated in [2–5]. The proposed method in this chapter is based on the Fourier transform that creates a representation of the signal in the frequency domain. The main points of the proposed method and the proposed technique are:

1. A Laplace transform is necessary to obtain convenient and simple input–output connections of the fields. The method consists of Fourier coefficients of the transverse dielectric profile and of the input–output profile. Thus, the accuracy of the method depends on the number of the modes in the system.

2. The Laplace transform

$$\tilde{a}(s) = L\{a(\zeta)\} = \int_{\zeta=0}^{\infty} a(\zeta)e^{-s\zeta}d\zeta, \tag{30}$$

is applied on the z-dimension, where $a(z)$ represents any z-dependent variables of the wave equations.

3. A Fourier transform is applied on the transverse dimension

$$\bar{g}(k_x, k_y) = F\{g(x,y)\} = \int_x\int_y g(x,y)e^{-jk_xx-jk_yy}dxdy, \tag{31}$$

and the differential equations are transformed to an algebraic form in the (ω, s, k_x, k_y) space.

4. The method of images is applied to satisfy the conditions $\hat{n} \times E = 0$ and $\hat{n} \cdot (\nabla \times E) = 0$ on the surface of the ideal metallic waveguide walls, where $\hat{n}$ is a unit vector perpendicular to the surface. The dielectric profile, $g(x,y)$, is defined inside the waveguide boundaries, $0 \le x \le a$ and $0 \le y \le b$. In order to maintain the boundary conditions without physical metallic walls, a substitute physical problem is constructed with infinite transverse extent. The periodicity and the symmetry properties are chosen to force the boundary conditions at the location of the walls in the real problem. This is done by extending the waveguide region $0 \le x \le a, \quad 0 \le y \le b$ to a four-fold larger region. Hence, the following relations are yielded

$$g(-x,y) = g(x,-y) = g(x,y) = g(-x,-y), \tag{32}$$

$$E_x(x,-y) = -E_x(x,y) \quad , \quad E_x(-x,y) = E_x(x,y). \tag{33}$$

The region $-a \le x \le a, \ -b \le y \le b$ is then further extended to infinity by periodic replication, $g(x + 2\ell a, y + 2kb) = g(x,y)$, where $-\infty < \ell, k < \infty$. The field components are periodically, namely, $E_x(x + 2\ell a, y + 2kb) = E_x(x,y)$ for $-\infty < \ell, k < \infty$. The substitution of the physical problem is equivalent to the original problem in the region $0 \le x \le a$, $0 \le y \le b$, and satisfies the same boundary conditions on the boundary of this region. The discrete Fourier transform series is given with $k_x = n\pi/a$ and $k_y = m\pi/b$, and the transverse wavenumbers are given by $k_{ox} = \pi/a$, and $k_{oy} = \pi/b$, where a and b are the transverse dimensions of the rectangular boundaries. We substitute $k_x = nk_{ox}$ and $k_y = mk_{oy}$, where the integers n and m are truncated by $-N \le n \le N$ and $-M \le m \le M$, respectively. The orders N and M determine the accuracy of the solution.

5. The output transverse field profiles are given by the inverse Laplace and Fourier transforms, as follows

$$\mathbf{E_y(x,y,z)} = \sum_{\mathbf{n}} \sum_{\mathbf{m}} \int_{\sigma - \mathbf{j}_\infty}^{\sigma + \mathbf{j}_\infty} \mathbf{E_y(n,m,s)} \exp\left[j\mathbf{nk_{ox}x} + \mathbf{jmk_{oy}y} + \mathbf{sz} \right] \mathbf{ds}, \tag{34}$$

where the inverse Laplace transform is calculated according to the Salzer method [17, 18].

6. The main objective is to generalize the mode model method [16] in order to solve also complicated and practical problems of circular dielectic material and a circular hollow core (**Figure 1(a)** and **(b)**) in the center of the cross section of the rectangular waveguide.

7. The second objective is to find the relevant parameters in order to obtain the Gaussian behavior of the output field in the interesting cases of circular dielectic material and a circular hollow core in the rectangular cross section.

Author details

Zion Menachem
Department of Electrical Engineering, Shamoon College of Engineering, Beer Sheva, Israel

*Address all correspondence to: zionme@sce.ac.il; zionmm@gmail.com

References

[1] Chiang KS. Review of numerical and approximate methods for the modal analysis of general optical dielectric waveguides. Optical and Quantum Electronics. 1994;**26**:S113-S134

[2] Wagner RL, Otto GP, Chew WC. Fast waveguide mode computation using wavelet-like basis functions. IEEE Microwave and Guided Wave Letters. 1993;**3**:208-210

[3] Goswami JC, Chan AK, Chui CK. An application of fast integral wavelet transform to waveguide mode identification. IEEE Transactions on Microwave Theory and Techniques. 1995;**43**:655-663

[4] Cammarata M, Rizzo P, Dutta D, Sohn H. Application of principal component analysis and wavelet transform to fatigue crack detection in waveguides. Smart Structures and Systems. 2010;**6**:349-362

[5] Sabetfakhri K, Katehi PBK. Analysis of integrated millimeter-wave and submillimeter-wave waveguides using orthonormal wavelet expansions. IEEE Transactions on Microwave Theory and Techniques. 1994;**42**:2412-2422

[6] Vaish A, Parthasarathy H. Analysis of a rectangular waveguide using finite element method. Progress in Electromagnetics Research C. 2008;**2**: 117-125

[7] Lugo DC, Wang J, Weller TM. Analytical and experiment study of multilayer dielectric rod waveguides. IEEE Transactions on Microwave Theory and Techniques. 2021;**69**:2088-2097

[8] Gao W, Lee WSL, Fujita M, Nagatsuma T, Fumeaux C, Withayachumnankul W. Characteristics of effective-medium-clad dielectric waveguides. IEEE Transactions on Terahertz Science and Technology. 2021; **11**:28-41

[9] Sharma J, De A. Full-wave analysis of dielectric rectangular waveguides. Progress In Electromagnetics Research M. 2010;**13**:121-131

[10] Yeap KH, Teh KH, Yeong KC, Lai KC, Loh MC. Propagation in dielectric rectangular waveguides. Optica Applicata. 2016;**XLVI**:317-330

[11] Rogge U, Pregla R. Method of lines for the analysis of dielectric waveguides. Journal of Lightwave Technology. 1993; **11**:2015-2020

[12] Smirnov YG, Smolkin YE. Investigation of the spectrum of the problem of normal waves in closed regular inhomogeneous dielectric waveguide of arbitrary cross section. Doklady Akademii Nauk. 2018;**478**: 627-630

[13] Dalarsson M, Nordebo S. TE-wave propagation in graded waveguide structures. Osa Continuum. 2020;**3**:67-76

[14] Srikanth KH. Studying the characteristics of a rectangular waveguide using HFSS. International Journal of Computer Applications (0975–8887). 2015;**118**(21):5-8

[15] Islamov IJ, Shukurov NM, Abdullayev RS, Hashimov KK, Khalilov AI. Diffraction of electromagnetic waves of rectangular waveguides with a longitudinal. IEEE. 2020;1–7

[16] Menachem Z, Jerby E. Transfer matrix function (TMF) for propagation

in dielectric waveguides with arbitrary transverse profiles. IEEE Transactions on Microwave Theory and Techniques. 1998;**46**:975-982

[17] Salzer HE. Orthogonal polynomials arising in the numerical evaluation of inverse Laplace transforms. Mathematical Tables and Other Aids to Computation. 1955;**9**:164-177

[18] Salzer HE. Additional formulas and tables for orthogonal polynomials originating from inversion integrals. Journal of Mathematical Physics. 1961; **39**:72-86

[19] The Numerical Algorithms Group (NAG) Ltd, Wilkinson House, Oxford, UK. 1999

[20] Collin RE. Foundation for Microwave Engineering. New York: McGraw-Hill; 1996